# Optical Furnaces for Crystal Growth

by
Gerhard Kloos

Well-defined single crystals of high quality are of importance in some branches of industry as well as in fundamental investigations of materials research. This monograph is devoted to the growth of crystals using optical furnaces. Optical furnaces can be understood and designed making recurrence to concepts that stem from analytical geometry. Therefore, these ideas are presented taking both "faces" of analytical geometry into account.

For the operation and maintenance of optical furnaces it is advantageous to gain an understanding of their principle of operation and alignment sensitivities. The method of analysis presented in this book strongly relies on pictorial representations and ray tracing is used as a means to visualize the working principles of these furnaces.

**Cover Graphic**

Infrared imaging furnace for crystal growth featuring two reflectors. Image courtesy of Quantum Design, Inc. © 2016.

# Optical Furnaces for Crystal Growth

by

**Gerhard Kloos**

Published by **Materials Research Forum LLC**
Millersville, PA 17551, USA

Published as part of the book series
**Materials Research Foundations**
Volume 9 (2017)
ISSN 2471-8890 (Print)
ISSN 2471-8904 (Online)

Print ISBN 978-1-945291-20-3
ePDF ISBN 978-1-945291-21-0

Distributed worldwide by

**Materials Research Forum LLC**
105 Springdale Lane
Millersville, PA 17551
USA
http://www.mrforum.com

Manufactured in the United State of America
10 9 8 7 6 5 4 3 2 1

# Table of Contents

*Preface*

Chapter 1    Crystal growth using optical heating ............................................. 1

Chapter 2    Reflector surfaces and shapes ...................................................... 14

Chapter 3    Imaging furnaces featuring direct illumination .......................... 30

Chapter 4    Imaging furnaces with intermediate focus................................... 77

Chapter 5    Laser heating ................................................................................. 88

*Keywords* ............................................................................................................ 99

*About the author* .............................................................................................. 101

# Preface

Well-defined single crystals of high quality are of importance in some branches of industry as well as in fundamental investigations of materials research. There is a wealth of methods to grow single crystals from an amorphous substance. This monograph is devoted to the growth of crystals using optical furnaces. These devices are applied to grow crystals of materials that have a high melting point.

Optical furnaces can be understood and designed making recurrence to concepts that stem from analytical geometry. Therefore, these ideas are presented taking both "faces" of analytical geometry into account, namely geometrical figures to illustrate reflector shapes on the one hand and formulas to describe mirror surfaces analytically on the other hand. Refinement of these basic geometries can then be realized in the design process using numerical optimization.

For the operation and maintenance of optical furnaces it is advantageous to gain an understanding of their principle of operation and alignment sensitivities. The method of analysis presented in this book strongly relies on pictorial representations and ray tracing is used as a means to visualize the working principles of these furnaces.

Gerhard Kloos

# CHAPTER 1

# Crystal growth using optical heating

**Abstract**

In this introductory text, main features of the crystal growth method using optical heating are presented. A first overview on some of the optical furnaces used for crystal growth is given. They have in common that optical systems based on mirrors are used to direct the light to the specimen and concentrate it in a focal region to obtain high temperatures. The optical floating-zone method is presented as a tool of materials research and a way to produce single crystals of high quality. We will have a look at some of the materials featuring a high melting point to which the method applies and on the form of the floating zone.

**Keywords**

Crystal-growth, growing techniques, optical heating, laser heating, compound mirror, floating-zone technique, melting points.

## Contents

| | | |
|---|---|---|
| 1. | Introduction | 2 |
| 2. | Growing Techniques | 2 |
| 3. | Optical heating | 3 |
| 3.1 | Optical heating with incoherent light sources | 3 |
| 3.2 | Laser heating | 6 |
| 4. | Materials | 7 |
| 5. | On the form of the floating zone | 10 |
| | References | 13 |

## 1.   Introduction

Unlike many other methods of crystal growth, the optical floating zone technique is based on optical heating, i.e. optical radiation is concentrated in a limited region to cause the material to melt. To this end, light sources that provide relatively high intensities are necessary as well as optical systems that focus the light in the region of the molten material.

Main advantages of this approach are the high temperatures that can be reached and that no crucible is necessary

## 2.   Growing techniques

Zone melting was proposed and developed by Pfann [1-3] as a means to purify materials. It turned out that it can also be used as a method to grow pure single crystals of high quality from substances that have a high melting point.

Using the optical floating-zone method, the material, which is to be melted, is brought into the focus of a reflector system and illuminated with high intensity by a halogen or xenon lamp. It is a salient feature of the method that the irradiated specimen is suspended.

Usually, an upper rod is mounted that consists of the amorphous material. A lower rod is mounted below, on which the single crystal will grow during the process. During crystal growth, a floating zone of molten material is formed between the two rods.

*Fig. 1.      Crystal-growth apparatus consisting of two reflectors that target the light on the floating zone inside a quartz tube. The set-up features an internal focus where the rays cross before impinging on the specimen.*

The crystal growth apparatus is generally equipped with a pulling mechanism. Its purpose is to perform a controlled vertical motion so that new amorphous material can enter the melting zone while crystallizing material can leave it and cool down.

It is also common practice to use two motors that can rotate the seed rod and the other rod independently. Depending on the experimental conditions, different speeds of rotation as well as different directions of rotation can be chosen. Choosing opposite directions of rotation for the upper and the lower rod results in increased mixing of the material in the melting zone.

It is an advantage of the crystal growth method that uses optical heating to induce melting in a suspended floating zone that no crucible is necessary. This avoids contamination problems and enhances the production of high-purity crystals.

## 3.    Optical heating

### 3.1    Optical heating with incoherent light sources

Traditionally, halogen lamps or xenon lamps are used as light sources to irradiate the melt zone with high intensity. There are several configurations to concentrate the light on the specimen. In one type of optical configuration, the light sources are  placed at one focal point of each ellipsoid while the specimen is situated at the conjugated focal point (Fig. 2). In another type, the light passes through an intermediate focus before the floating zone is reached (Fig. 3).

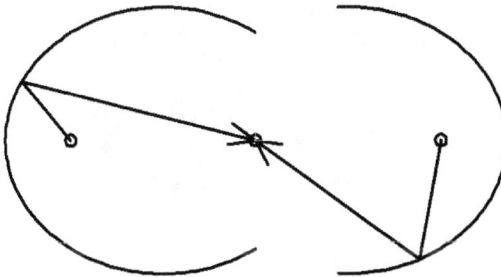

*Fig. 2.        Direct illumination of the melting zone using two reflectors.*

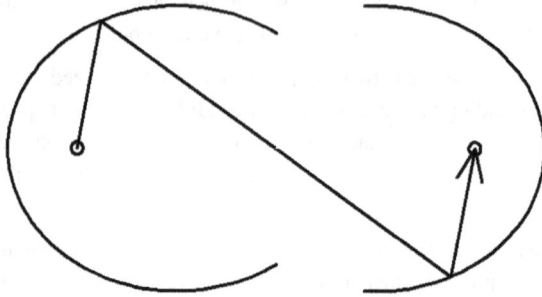

*Fig. 3.* *Illumination of the melting zone using an optical arrangement with intermediate focus.*

The optical reflectors that are used to concentrate the light may be closed (Fig. 4) or open Fig. 5).

*Fig. 4.* *Closed double ellipsoid.*

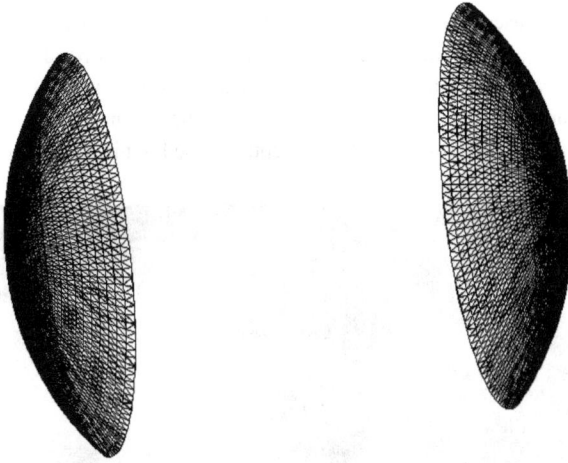

*Fig. 5.*      *Two separate partial ellipsoidal mirrors.*

Open reflectors have the advantage of better accessibility while closed mirrors provide higher optical efficiency.

## 3.2    Laser heating

The laser-heated pedestal growth method [6] makes use of a coherent light source.

The energy of the laser is again concentrated in a focal region to melt the material, but the optical system is quite different, because the laser beam is quasi-parallel. In a first optical unit, the laser beam is expanded and a circular pattern of rays is created. This ray fan is directed onto an annular mirror, from which it is reflected in the direction of a partial paraboloidal mirror. This mirror finally concentrates the light in the melting region.

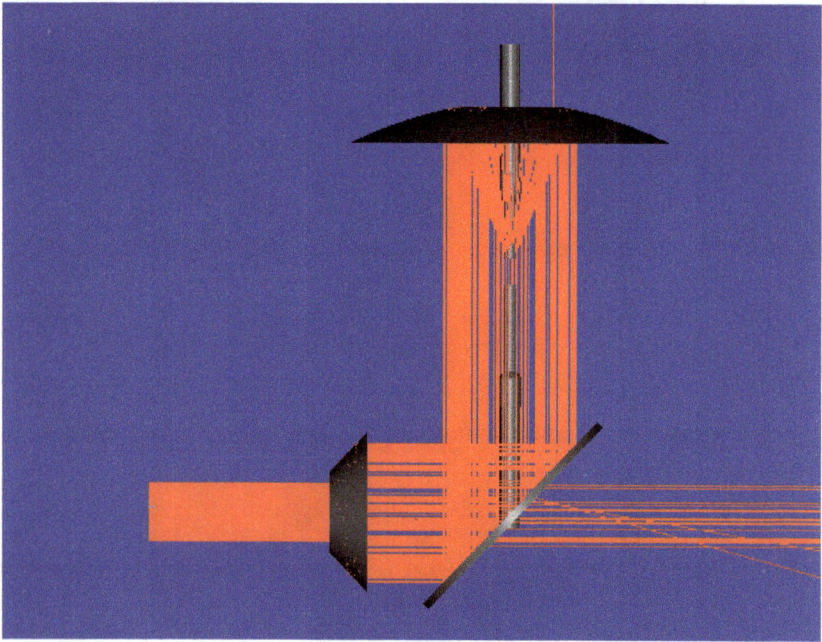

*Fig. 6.        Crystal-growth apparatus using a laser to heat the sample.*

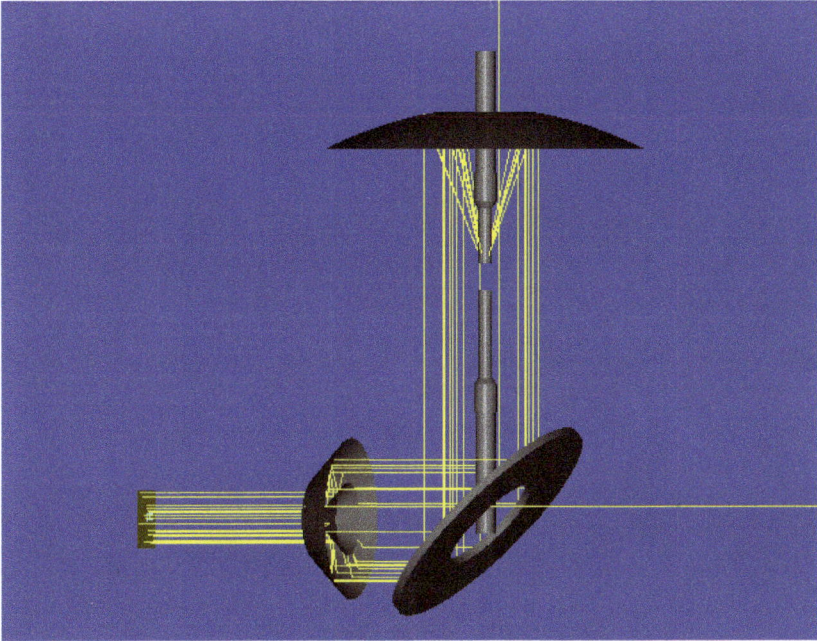

*Fig. 7.*     *Laser heating: The rays are targeted at the tip of the upper rod.*

## 4.    Materials

Optical zone-melting is applied to obtain single crystals of substances with high melting points and engineering work has been devoted to extend the temperature range that can be covered by a given optical oven.

*Table 1.* *Approximate melting temperatures of some orthoferrites*

| Material | Melting point [°C] |
|---|---|
| | 1680 |
| $YbFeO_3$ | 1690 |
| $YFeO_3$ | 1700 |
| $TbFeO_3$ | 1720 |
| $LaFeO_3$ | 1920 |

Orthoferrites are chemical compounds with the chemical formula $RFeO_3$ where the symbol R stands for one or more rare-earth elements. These compounds feature an orthorhombic crystal structure. They show large magneto-optical effects.

*Table 2.* *Melting temperatures of some materials*

| Material | Melting point [°C] |
|---|---|
| GaSb | 517 |
| Al | 660 |
| $LiNbO_3$ | 1280 |
| Si | 1410 |
| $Al_2O_3$ | 2072 |
| NiO | 2087 |

Lithium niobate ($LiNbO_3$) is an interesting material from the technical point of view. The material displays ferroelectricity, the Pockels effect, piezoelectricity, and special nonlinear optical properties. It is used in optical waveguides and piezoelectric sensors as well as in optical modulators. There are also applications in the field of laser engineering such as frequency doubling and so-called Q-switching.

*Table 3.*    *Melting temperature of some titanium compounds*

| Material | Melting point [°C] |
|---|---|
| BaTiO$_3$ | 1625 |
| TiO$_2$ | 1843 |
| TiC | 3160 |

The inorganic compound barium titanate (BaTiO$_3$) is a photorefractive material that is used in applications of nonlinear optics. In addition, the piezoelectric properties of this material are of technological interest. During the growth of single crystals of this compound an experimental difficulty can be encountered, because the material has a phase transition from hexagonal to cubic phases at 1460 °C [7].

An interesting account on the application of the method to the crystal growth of oxides is given in reference [5]. It also provides experimental details on the characterization of the crystals that are grown.

Before the process of producing a single crystal can be started, a feeding rod has to be prepared. To this end, a powder is prepared and mixed in the mortar. After annealing, a rod is formed from the material using cold or hot pressing. Usually, a sintering step with carefully controlled temperature program and gas atmosphere in the oven follows then.

To control the onset of crystallization, it is advantageous to use a single crystal as seed and mount it on the lower rod before the growth process is started.

Using crystal growth by optical heating, high-quality crystals with well-defined properties can be obtained.

From the point of view of materials research, the floating zone method is also a precious tool to investigate phase diagrams [8].

The advantage of the floating zone method that no crucible is being used allows for the preparation of solid solutions with controlled and uniform chemical composition [5].

## 5.    On the form of the floating zone

Exposed to the intense optical radiation, the substance in the focal region is melted and behaves as a highly viscous fluid. The external shape of the melting zone is influenced by gravity and takes a form similar to the one sketched in Fig. 8.

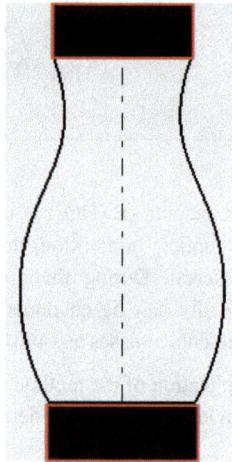

Fig. 8.        *Molten substance between the two rods.*

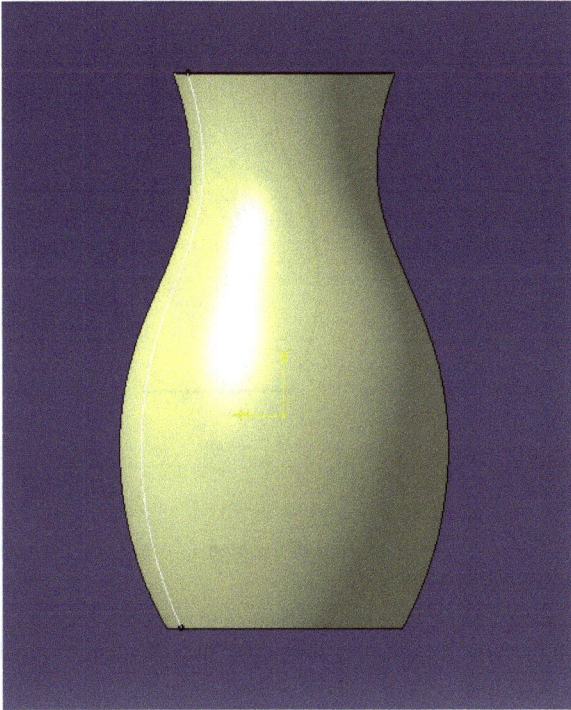

*Fig. 9.      Shape of the floating zone between the two rods*

There are different ways to describe the shape of the floating zone:

Locally, radii of curvature can be attributed to each point of the surface, i.e. the interface between liquid and air. Their values depend on the local surface tensions.

The form of the melting zone may also be described macroscopically by stating the contact angles at the upper and at the lower rod.

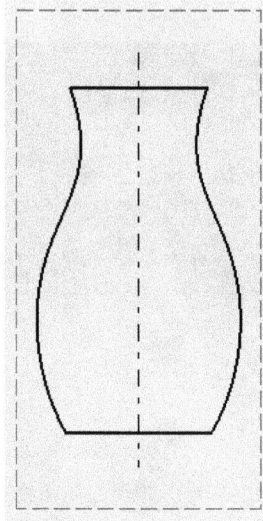

*Fig. 10.      Axis of symmetry (z-axis) of the body of rotation, by which the floating zone might be described approximately.*

In many cases, the floating zone can be assumed to be symmetric with respect to the axis. Let $f(z)$ be a function that describes the surface of the melting material in the x-z-plane. The melting zone might then be approximated as a figure of revolution, the radius of which depends on the height $z$ and changes with height as $f(z)$.

$$\begin{bmatrix} x \\ y \\ z \end{bmatrix} = \begin{bmatrix} f(z)\cos\varphi \\ f(z)\sin\varphi \\ z \end{bmatrix} \tag{1}$$

Many details on the description and interpretation of the shape of molten substances in crystal growth may be found in a monograph on the subject [9].

**References**

[1] W.G. Pfann, Principles of zone melting, Trans. Am. Inst. Mining Metall. Eng. 194 (1952) 747-753.

[2] W. Pfann, Zone melting, Science 135(3509) (1962) 1101-1109. https://doi.org/10.1126/science.135.3509.1101

[3] W.G. Pfann, Zone Melting, third ed., Robert E. Krieger Pub. Co., Huntington, New York, 1978.

[4] P.E. Glaser, Imaging-furnace developments for high-temperature research, J. of the Electrochemical Society 107 (1960) 226-231. https://doi.org/10.1149/1.2427656

[5] H.A. Dabkowska, A.B. Dabkowski, Crystal Growth of Oxides by Optical Floating Zone Technique, in: G. Dhanaraj, K. Byrappa, V. Prasad, M. Dudley (Eds.), Springer Handbook of Crystal Growth, Springer-Verlag, Berlin, Heidelberg, 2010, pp. 367-391. https://doi.org/10.1007/978-3-540-74761-1_12

[6] M.R.B. Andreeta, A.C. Hernandes, Laser-Heated Pedestal Growth of Oxide Fibers, in: G. Dhanaraj, K. Byrappa, V. Prasad, M. Dudley (Eds.), Springer Handbook of Crystal Growth, Springer-Verlag, Berlin, Heidelberg, 2010, pp. 367-391. https://doi.org/10.1007/978-3-540-74761-1_13

[7] C.W. Lan, J.C. Leu, Y. Huang, On the design of double-ellipsoid mirror furnace and its thermal characteristics for floating-zone growth of $Sr_xBa_{1-x}TiO_2$ single crystals, Cryst. Res. Technol. 35 (2000) 167-176. https://doi.org/10.1002/(SICI)1521-4079(200002)35:2<167::AID-CRAT167>3.0.CO;2-9

[8] A. Miyazaki, H. Kimura, X. Jia, Crystal and glass growth in $BaO-B_2O_3-Al_2O_3$ system by floating zone pulling down method, Cryst. Res. Technol. 35 (2000) 1245-1250. https://doi.org/10.1002/1521-4079(200011)35:11/12<1245::AID-CRAT1245>3.0.CO;2-R

[9] V.A. Tatartchenko, Shaped Crystal Growth, in: G. Dhanaraj, K. Byrappa, V. Prasad, M. Dudley (Eds.), Springer Handbook of Crystal Growth, Springer-Verlag, Berlin, Heidelberg, 2010, pp. 367-391. https://doi.org/10.1007/978-3-540-74761-1_16

# CHAPTER 2

# Reflector surfaces and shapes

**Abstract**

The geometrical and optical properties of ellipsoids and reflecting surfaces derived from ellipsoids form the basis of imaging furnaces used for crystal growth. The geometry of ellipses is used as starting point to introduce ellipsoids. The method due to de la Hire is presented as a simple way to construct ellipses geometrically. Rotation-symmetric ellipsoids as well as three-axial ellipsoids are considered. The influence of scaling operations on the focal points of an anamorphic ellipsoid is discussed. It is shown how conics in general can be described and the discussion is extended to aspheric surfaces in the broader sense. Finally, an equation is derived that can be used to calculate the ray reflected by a given mirror.

**Keywords**

Ellipses, ellipsoids, reflectors, conics, aspherical surfaces.

## Contents

1.  Introduction.................................................................................................15

2.  Ellipsoidal geometries...............................................................................15

2.1  Ellipses .....................................................................................................15

2.2  Describing ellipsoidal surfaces ...............................................................22

2.3  Describing conic surfaces.........................................................................25

2.3  Describing aspherical surfaces ................................................................26

2.4  Tracing rays ..............................................................................................27

References .........................................................................................................29

## 1. Introduction

Ellipsoidal mirrors might be considered as the workhorse of many optical furnaces used for crystal growth. This is due to the fact that they have the property of being able to concentrate the energy of a light source in a limited space, where the polycrystalline sample will be located. It is therefore useful to have a closer look at their geometrical and optical properties and how these properties change if the form of the optical reflector is changed.

## 2. Ellipsoidal geometries

### 2.1 Ellipses

As a prerequisite to considering three-dimensional objects (reflecting surfaces), let us first have a look at some properties of an ellipse. Geometrical relations will be stated that are useful for understanding as well as for designing reflectors for optical heating.

A nice way to find the ellipse that is described by a given semi-major axis a and by a given semi-minor axis b is the geometrical *construction of de la Hire*. It starts by representing both axes by concentric circles.

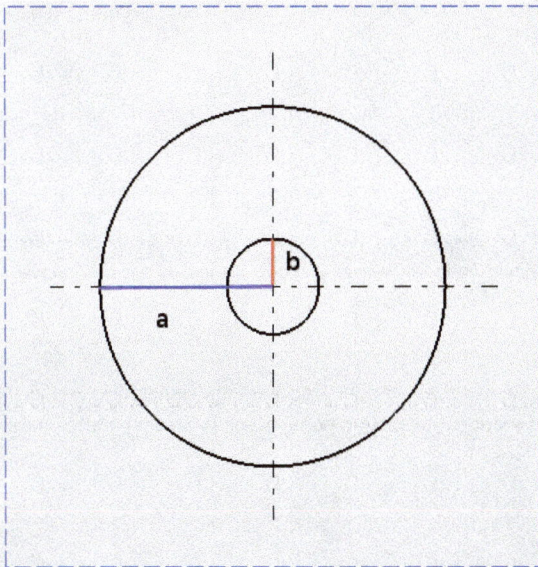

*Fig. 1.*     *On the geometrical construction due to de la Hire. In a first step, two circles are drawn that correspond to the semi-minor and semi-major axis, respectively.*

A straight line is drawn then that starts at the origin of the coordinate system and one of its intersection points with the outer circle is determined. From this intersection point, a line is drawn that is parallel to the y-axis (y-line).

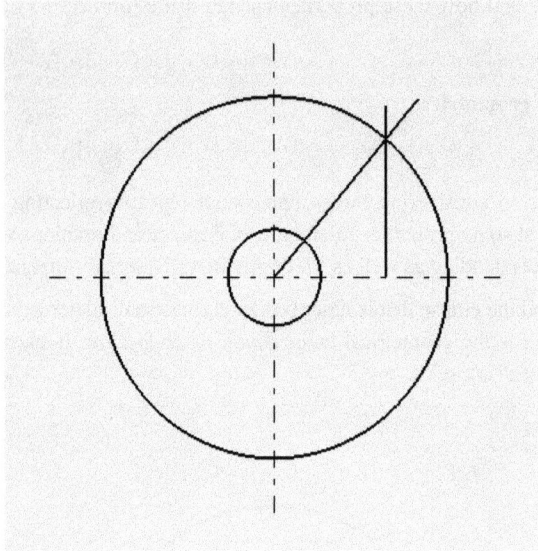

*Fig. 2.* *On the geometrical construction due to de la Hire. In a second step, a straight line is drawn from the center of both circles. From the intersection point on the outer circle, a vertical line is drawn.*

The intersection of the first straight line with the inner circle is determined as well. Starting from this second intersection point, a line is drawn parallel to the x-axis (x-line).

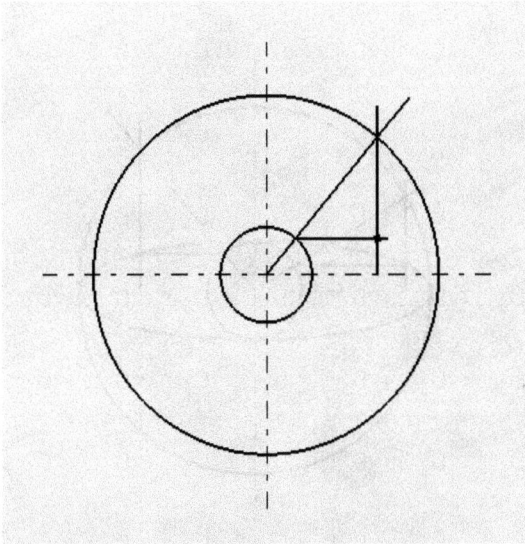

*Fig. 3.* *On the geometrical construction due to de la Hire. In a third step, a horizontal straight line is draw from the intersection point on the inner circle and the intersection point with the vertical line is determined.*

The point where x-line and y-line intersect is a point on the ellipse.

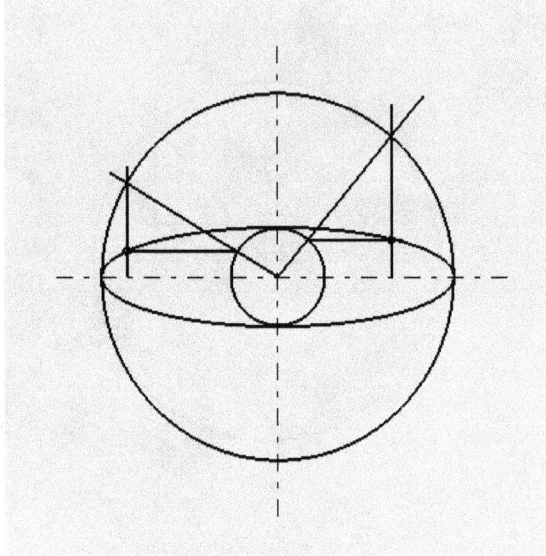

*Fig. 4.        On the geometrical construction due to de la Hire. In this way, other points
on the ellipse can be found,*

To understand why this is the case, this procedure may be described analytically. The
outer circle has the equation

$$x^2 + y^2 = a^2 \tag{1}$$

while the inner circle has the equation

$$x^2 + y^2 = b^2 \tag{2}$$

This implies that the y-line, which starts from the outer circle, has the form

$$\begin{bmatrix} x \\ y \end{bmatrix} = a \begin{bmatrix} \cos\varphi \\ \sin\varphi \end{bmatrix} + s \begin{bmatrix} 0 \\ 1 \end{bmatrix} \tag{3}$$

while the x-line, which starts from the inner circle, takes the form

$$\begin{bmatrix} x \\ y \end{bmatrix} = b \begin{bmatrix} \cos\varphi \\ \sin\varphi \end{bmatrix} + t \begin{bmatrix} 1 \\ 0 \end{bmatrix} \tag{4}$$

wherein s and t are parameters, which have to be determined. The intersection point can be calculated as the locus that both lines have in common:

$$s\begin{bmatrix} 0 \\ 1 \end{bmatrix} - t\begin{bmatrix} 1 \\ 0 \end{bmatrix} = -a\begin{bmatrix} \cos\varphi \\ \sin\varphi \end{bmatrix} + b\begin{bmatrix} \cos\varphi \\ \sin\varphi \end{bmatrix} \tag{5}$$

As solution, we obtain the following expressions for the parameters s and t in terms of the axes of the ellipse and of the angle of the first line with respect to the x-axis.

$$s = +(b-a)\sin\varphi \tag{6}$$

$$t = -(b-a)\cos\varphi \tag{7}$$

To calculate the intersection point of both lines, we may either introduce the parameter s into Eq. 3 or introduce the parameter t into Eq. 4. In both cases, we obtain as result for the coordinates of the intersection point:

$$\begin{bmatrix} x \\ y \end{bmatrix} = \begin{bmatrix} a\cos\varphi \\ b\sin\varphi \end{bmatrix} \tag{8}$$

This point lays on the ellipse with semi-major axis a and semi-minor axis b. In fact, Eq. 8 is a parametric form of the equation of the ellipse.

Squaring the components of the vector, we obtain the implicit form of the equation of an ellipse [1] that has the origin of the coordinate system at its center:

$$\frac{x^2}{a^2} + \frac{y^2}{b^2} = 1 \tag{9}$$

The ellipse has *two focal points*, the position of which is given by a quadratic equation:

$$f_1^2 = a^2 - b^2 \tag{10}$$

Even though the geometrical construction of de la Hire makes no use of the concept of a focal point, we can retrieve the focal points after having obtained the shape of the ellipse from this geometrical construction. To this end, we shift the outer circle with radius a in the vertical direction, i.e. along the semi-minor axis of the ellipse by a distance b. The center of the outer circle now coincides with the ellipse. The points where the shifted outer circle and the x-axis intersect are the focal points of the ellipse.

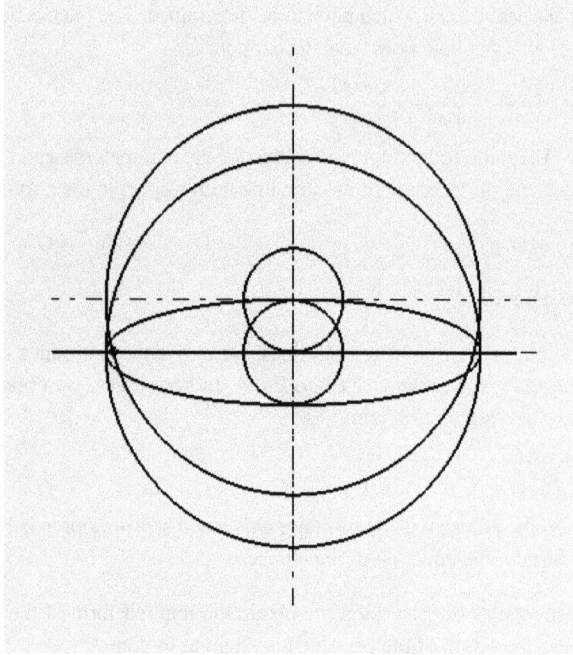

*Fig. 5.* Finding the focal point of the ellipse. The outer circle is translated by a distance that corresponds to the radius of the inner circle along the vertical axis. The intersection points of this shifted circle with the horizontal axis of the coordinate system are the positions of the focal points.

For optical considerations, the *normal* of the ellipse at a given point $(x_0, y_0)$ is of importance. The normal can be determined as gradient of the function

$$f(x, y) = \frac{x^2}{a^2} + \frac{y^2}{b^2} - 1 \tag{11}$$

It reads

$$\nabla f = \begin{bmatrix} \dfrac{2}{a^2} x \\ \dfrac{2}{b^2} y \end{bmatrix} \tag{12}$$

To find the equation of the normal at the point $(x_0, y_0)$, we can first state the linear equation as follows:

$$\vec{N}\vec{x} = g \tag{13}$$

where

$\vec{N} = 2\begin{bmatrix} \dfrac{x_0}{a^2} \\ \dfrac{y_0}{b^2} \end{bmatrix}$ is the normal vector at the point $(x_0, y_0)$ and g is the constant that has

to be determined. Its value follows directly by using the condition that the point $(x_0, y_0)$ is a point on the ellipse. Introducing these coordinates into the above equation, we have

$$g = 2\left(\frac{x_0^2}{a^2} + \frac{y_0^2}{b^2}\right) \tag{14}$$

The equation of the normal of the ellipse at this point therefore reads

$$\frac{x_0 x}{a^2} + \frac{y_0 y}{b^2} = 1 \tag{15}$$

This is an equation in implicit form. The corresponding equation in parametric form is

$$\vec{x} = \begin{bmatrix} x_0 \\ y_0 \end{bmatrix} + \lambda \vec{N} \tag{16}$$

wherein the parameter $\lambda$ is a real number.

For completeness, we might also consider the *tangent* to the ellipse at the same point. The tangent has the property of being perpendicular to the normal. To find its equation, we first state a vector that is orthogonal to the normal vector of the normal.

$$\vec{d} \perp \nabla f \tag{17}$$

$\vec{d} = \begin{bmatrix} -\dfrac{y}{b^2} \\ \dfrac{x}{a^2} \end{bmatrix}$ is such a vector. We can therefore write down the equation of the tangent at the given point in the following form:

$$\vec{T}\vec{x} = h \tag{18}$$

In this linear equation the vector

$$\vec{T} = \begin{bmatrix} -\dfrac{y_0}{b^2} \\ \dfrac{x_0}{a^2} \end{bmatrix} \tag{19}$$

appears and h has to be determined. Making use of the condition that the point $(x_0, y_0)$ lays on the ellipse, we have

$$h = \left( \frac{1}{a^2} - \frac{1}{b^2} \right) x_0 y_0 \tag{20}$$

This implies that the equation of the tangent at this point takes the following form:

$$\begin{bmatrix} -\dfrac{y_0}{b^2} \\ \dfrac{x_0}{a^2} \end{bmatrix} \overrightarrow{x} = \left( \frac{1}{a^2} - \frac{1}{b^2} \right) x_0 y_0 \tag{21}$$

Again, we stated the implicit form of the equation first. The parametric form of the equation of the tangent is

$$\overrightarrow{x} = \begin{bmatrix} x_0 \\ y_0 \end{bmatrix} + \lambda \overrightarrow{T} \tag{22}$$

with the real number $\lambda$ as parameter of the linear equation.

## 2.2 Describing ellipsoidal surfaces

Ellipsoidal mirrors are a key component of many optical furnaces. This is due to the fact that their form allows to concentrate light.

If we rotate the ellipse about its axis, we obtain an ellipsoid that is symmetric with respect to that axis

$$\frac{x^2}{a^2} + \frac{y^2}{a^2} + \frac{z^2}{c^2} = 1 \tag{23}$$

A common way to describe ellipsoidal surfaces analytically is the implicit form of the general equation of the ellipsoid.

$$\frac{x^2}{a^2} + \frac{y^2}{b^2} + \frac{z^2}{c^2} = 1 \tag{24}$$

This equation refers to an ellipsoid that has its center positioned at the origin of the coordinate system. $a$, $b$, and $c$ describe the length of the ellipsoid along the x-, y-, and z-axis, respectively.

The focal lengths of the ellipsoid in the x-y-, x-z-, and y-z-plane, respectively are

$$f_1^2 = a^2 - b^2 \tag{25}$$

$$f_2^2 = a^2 - c^2 \tag{26}$$

$$f_3^2 = b^2 - c^2 \tag{27}$$

There are several ways to scale ellipsoidal surfaces to tailor them for a design task.

One way of scaling (*scaling method A*) is the following, in which a parameter $\lambda$, which has the dimension of the square of a length, is introduced into the standard equation.

$$\frac{x^2}{a^2 + \lambda} + \frac{y^2}{b^2 + \lambda} + \frac{z^2}{c^2 + \lambda} = 1 \tag{28}$$

The parameter takes values other than zero, because this case would correspond to no scaling.

The new ellipsoids created by this scaling operation have the interesting property of being confocal to the original ellipsoid. To see this, it might be helpful to introduce abbreviations for the lengths of the new ellipsoids:

$$A^2 = a^2 + \lambda \tag{29}$$

$$B^2 = b^2 + \lambda \tag{30}$$

$$C^2 = c^2 + \lambda \tag{31}$$

If we now calculate the focal lengths of the scaled ellipsoid as we did before

$$F_1^2 = A^2 - B^2 \tag{32}$$

$$F_2^2 = A^2 - C^2 \tag{33}$$

$$F_3^2 = B^2 - C^2 \tag{34}$$

we find

$$F_1^2 = \left(a^2 + \lambda\right) - \left(b^2 + \lambda\right) = f_1^2 \tag{35}$$

$$F_2^2 = \left(a^2 + \lambda\right) - \left(c^2 + \lambda\right) = f_1^2 \tag{36}$$

$$F_3^2 = \left(b^2 + \lambda\right) - \left(c^2 + \lambda\right) = f_3^2 \tag{37}$$

The focal lengths are the same. The ellipsoid scaled by the parameter $\lambda$ is therefore confocal to the ellipsoid that is described by Eq. 1.

Alternatively, the following method of scaling (*scaling method B*) may be used that leads to different results. Here, a parameter is introduced in the right-hand side of Eq. 1.

$$\frac{x^2}{a^2} + \frac{y^2}{b^2} + \frac{z^2}{c^2} = k^2 \tag{38}$$

This parameter $k$ takes values that are not equal to 1 nor to -1. As before, we introduce abbreviations for the axes of the scaled ellipsoid to see how the corresponding focal lengths will change:

$$A^2 = ka^2 \tag{39}$$

$$B^2 = kb^2 \tag{40}$$

$$C^2 = kc^2 \tag{41}$$

We use Eqs. 32-34 again, but the symbols now refer to the alternative way of scaling the ellipsoid.

$$F_1^2 = k^2\left(a^2 - b^2\right) \neq f_1^2 \tag{42}$$

$$F_2^2 = k^2\left(a^2 - c^2\right) \neq f_2^2 \tag{43}$$

$$F_3^2 = k^2\left(b^2 - c^2\right) \neq f_3^2 \tag{44}$$

This method of scaling differs significantly from the first one: The scaled ellipsoid is not confocal with respect to the original one.

Depending on the task, it might be appropriate to use *lateral scaling*.

If we apply scaling in a similar way as in method A, but restrict the scaling to the first and to the second axis of the ellipsoid, we have

$$\frac{x^2}{a^2 + \lambda} + \frac{y^2}{b^2 + \lambda} + \frac{z^2}{c^2} = 1 \tag{45}$$

The effect on the focal lengths is quite different:

$$F_1^2 = \left(a^2 + \lambda\right) - \left(b^2 + \lambda\right) = f_1^2 \tag{46}$$

$$F_2^2 = \left(a^2 + \lambda\right) - c^2 = f_1^2 + \lambda \tag{47}$$

$$F_3^2 = \left(b^2 + \lambda\right) - c^2 = f_3^2 + \lambda \tag{48}$$

The lateral scaling approach similar to method B might be stated as

$$\frac{x^2}{k^2a^2} + \frac{y^2}{k^2b^2} + \frac{z^2}{c^2} = 1 \tag{49}$$

Here, we have the following relations for the foci of the anamorphic ellipsoidal surface:

$$F_1^2 = k^2\left(a^2 - b^2\right) = k^2 f_1^2 \tag{50}$$

$$F_2^2 = k^2a^2 - c^2 \tag{51}$$

$$F_3^2 = k^2b^2 - c^2 \tag{52}$$

## 2.3    Describing conic surfaces

In the preceding paragraph, ellipsoidal reflectors were treated. In this paragraph, we will have a look at conic surfaces that comprise ellipsoidal surfaces as a special case. The description will be restricted to surfaces that are symmetric with respect to one axis.

Conic surfaces can be described as deviation from a sphere. It is therefore convenient to start with the equation of a sphere with radius R:

$$x^2 + y^2 + (z - R)^2 = R^2 \tag{53}$$

This sphere is positioned at the point $(x, y, z) = (0, 0, R)$. The purpose of this choice is that the apex of the reflectors, which will be described, coincides with the origin of the coordinate system.

To extend the description to conic mirrors in general, it is sufficient to introduce the conic constant k into the above equation:

$$x^2 + y^2 + (z - R)^2 + kz^2 = R^2 \tag{54}$$

This constant is an appropriate measure for the deviation of the reflector surface from the form of a sphere. And it allows for a useful classification of conical surfaces:

Table 1: Classification of conic reflectors using the conic constant as parameter

| $k < -1$ | $k = 1$ | $-1 < k < 0$ | $k = 0$ | $k > 0$ |
|---|---|---|---|---|
| Hyperboloid | Paraboloid | Prolate Ellipsoid | Sphere | Oblate Ellipsoid |

The equation above is the implicit form of the equation of a conic as it is used in textbooks on analytical geometry. The implicit form of this equation is frequently used in manuals on optical design programs. For this reason, it will be shown here how both equations are related. Let us first introduce the curvature $\kappa$ as another way to express the radius of curvature:

$$R = \frac{1}{\kappa} \tag{55}$$

Multiplication with $\kappa$ leads to

$$-k\kappa z^2 - \kappa\left(x^2 + y^2 + z^2\right) + 2z = 0 \tag{56}$$

This equation can now be re-arranged a little bit.

$$-(1+k)\kappa z^2 - \kappa\left(x^2 + y^2\right) + 2z = 0 \tag{57}$$

We now multiply both sides of the equation by $x^2 + y^2$ and add $z^2$ as well. In this way, the following equation is obtained:

$$z^2 - (1+k)\kappa^2\left(x^2 + y^2\right)z^2 = \kappa^2\left(x^2 + y^2\right)^2 - 2\kappa\left(x^2 + y^2\right)z + z^2$$

Then, we might re-arrange this equation to have a more concise form.

$$\left(1 - (1+k)\kappa^2\left(x^2 + y^2\right)\right)z^2 = \left(\kappa\left(x^2 + y^2\right) - z\right)^2 \tag{59}$$

We are targeting at an explicit equation with the variable $z$ on the left-hand side. To this end, we make the next step:

$$z + \sqrt{1 - (1+k)\kappa^2\left(x^2 + y^2\right)}\,z = \kappa\left(x^2 + y^2\right) \tag{60}$$

This implies that the following equation holds:

$$z = \frac{\kappa\left(x^2 + y^2\right)}{1 + \sqrt{1 - (1+k)\kappa^2\left(x^2 + y^2\right)}} \tag{61}$$

This is an explicit form of the conic equation, which describes the mirror surface in terms of the local curvature $\kappa$ and the conic constant $k$.

## 2.4 Describing aspherical surfaces

A more general class of mirror surfaces can be treated and optimized using computer programs by making recurrence to aspherical coefficients. They provide additional variables for a numerical optimization of a reflector for a given illumination task.

Due to the symmetry of the equation, it seems appropriate to put

$$r_\perp^2 = x^2 + y^2 \tag{62}$$

The quantity corresponds to the distance from the optical axis and the above equation takes the form

$$z = \frac{\kappa r_\perp^2}{1 + \sqrt{1 - (1+k)\kappa^2 r_\perp^2}} \tag{63}$$

This analytical equation is now extended in an ad-hoc way by adding an even polynomial

$$\sum_{l=1}^{N} a_l r_\perp^{2l} \tag{64}$$

The final equation is a combination of the analytical part, which describes a conic surface, and a polynomial expansion.

$$z = \frac{\kappa r_\perp^2}{1 + \sqrt{1 - (1+k)\kappa^2 r_\perp^2}} + a_1 r_\perp^2 + a_2 r_\perp^4 + a_3 r_\perp^6 + a_4 r_\perp^8 + \dots \tag{65}$$

The coefficients that appear on the right-hand side are called aspheric coefficients and they bear information on the form of the mirror surface.

(There is a certain redundancy between the conic constant and the aspheric constant $a_1$ as can be seen by Taylor expansion. For this reason, this aspheric constant is omitted in some optics software.)

## 2.5   Tracing rays

The law of reflection in vector form is a good starting point to trace rays that are reflected from the surface of a mirror at a point P.

$$\vec{u} = \vec{\imath} - 2\left(\hat{N} \cdot \vec{\imath}\right) \cdot \hat{N} \tag{66}$$

In this equation, $\vec{\imath}$ is the vector of the incident ray, i.e. of the ray that impinges on the mirror. $\hat{N}$ is the normal of the reflector at the point P and $\vec{u}$ is the vector of the reflected ray at this point of the mirror.

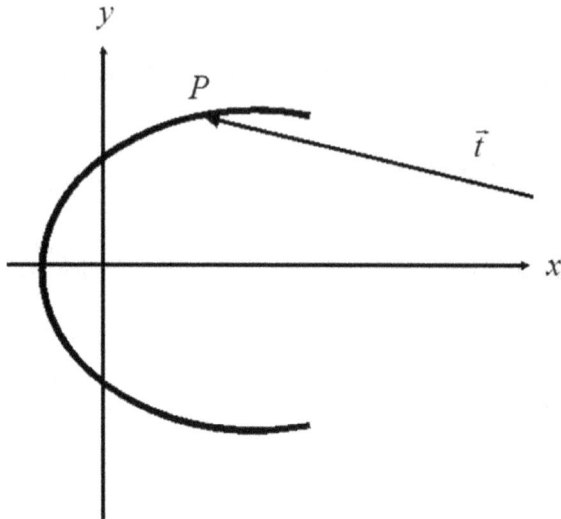

*Fig. 6.*     *x-y-cut through an optical reflector. A ray of light is reflected at a point P.*

For the sake of simplicity, we will restrict the discussion to a two-dimensional cut through the optical reflector that can be described using a function $y = f(x)$. To proceed, the local slope at the point P of the mirror is an important quantity.

$$m := f'(x)\big|_{P_1} \tag{67}$$

The corresponding tangent vector at this point is

$$\vec{T} = \begin{bmatrix} 1 \\ m \end{bmatrix} \tag{68}$$

The normal vector must be perpendicular with respect to the tangent vector.

$$\vec{T} \cdot \vec{N} = 0 \tag{69}$$

To determine it, we make use of a test function:

$$\vec{N} = \begin{bmatrix} -1 \\ g(p_1) \end{bmatrix} \tag{70}$$

It follows that

$$g(p_1) = \frac{1}{m} = \left( f'(x) \big|_{p_1} \right)^{-1} \tag{71}$$

This leads to the following result for the normal vector:

$$\vec{N} = \begin{bmatrix} -1 \\ 1/m \end{bmatrix} \tag{72}$$

Accordingly, the normalized normal vector reads

$$\hat{N} = \frac{m}{\sqrt{1+m^2}} \begin{bmatrix} -1 \\ 1/m \end{bmatrix} \tag{1}$$

We may now introduce the results into the equation with which we started.

$$\vec{u} = \vec{t} - \frac{2m}{\sqrt{1+m^2}} \left( \begin{bmatrix} -1 \\ 1/m \end{bmatrix} \cdot \begin{bmatrix} t_1 \\ t_2 \end{bmatrix} \right) \frac{m}{\sqrt{1+m^2}} \begin{bmatrix} -1 \\ 1/m \end{bmatrix} \tag{73}$$

This equation might be slightly re-arranged.

$$\vec{u} = \vec{t} - \frac{2m^2}{1+m^2} \left( -t_1 + \frac{1}{m} t_2 \right) \begin{bmatrix} -1 \\ 1/m \end{bmatrix} \tag{74}$$

In this way, an equation has been found that links the ray that impinges on a mirror at a given point to the ray that is reflected at this point. The mirror itself is described by its local slope, which can be calculated from the function that describes the surface of this reflector.

Even though we considered a simple situation, the mathematical description is relatively complicated and tracing a multitude of rays in this way would be tedious. Therefore, recurrence to computer programs is often made to solve problems of reflector design.

## References

[1]   B. Spain, Analytical Conics, Dover Publications, Mineola, New York, 2007. (unabridged republication of the work originally published in 1957 by Pergamon Press, London and New York)

[2]   H.F. Brueggeman, Conic Mirrors, The Focal Press, London, New York, 1968.

# CHAPTER 3

# Imaging furnaces featuring direct illumination

**Abstract**

Experimental set-ups for crystal growth are considered that are based on direct illumination of the melting zone using one or more ellipsoidal optical reflectors. The systems are discussed in the order of increasing complexity. To familiarize with the salient features of the illumination technique, closed as well as open single ellipsoid mirrors are considered first. Then, experimental configurations are described that use compound mirrors, which are made up of more than one ellipsoidal reflector, to illuminate the sample and obtain high temperatures inside the melting zone.

Ray-tracing and representations of caustics are used to visualize the ray paths inside the apparatus used for crystal growth.

**Keywords**

Crystal-growth apparatus, mono-ellipsoid, multiple ellipsoids, closed reflector, open reflector, compound mirror, ray tracing, caustics.

## Contents

1.    Introduction ................................................................................................31

2.    Mono-ellipsoidal crystal-growth apparatus .........................................31

2.1   Closed reflectors .........................................................................................31

2.2   Open reflectors ...........................................................................................43

3.    Crystal-growth apparatus using multiple ellipsoids ...........................69

References ...............................................................................................................76

## 1. Introduction

There is a variety of experimental configurations that use direct illumination of the melting zone. Installations are applied that use one, two, four or six ellipsoidal mirrors to illuminate the zone between the polycrystalline feeding rod and the grown crystal. It seems to be a natural way to start with simple configurations and then to turn to compound systems.

Many optical features of the different mirror configurations can be understood by visualizing rays passing through the device. These considerations give information that can be used for design purposes as well as during adjustment work.

## 2. Mono-ellipsoidal crystal-growth apparatus

### 2.1 Closed reflectors

Let us consider the experimental set-up for crystal growth depicted in Fig. 1. A halogen lamp serves as the source of energy. It illuminates the floating zone between a polycrystalline rod (above) and a growing single-crystal rod (below). Both rods are situated in a tube made of quartz, which can be pressurized.

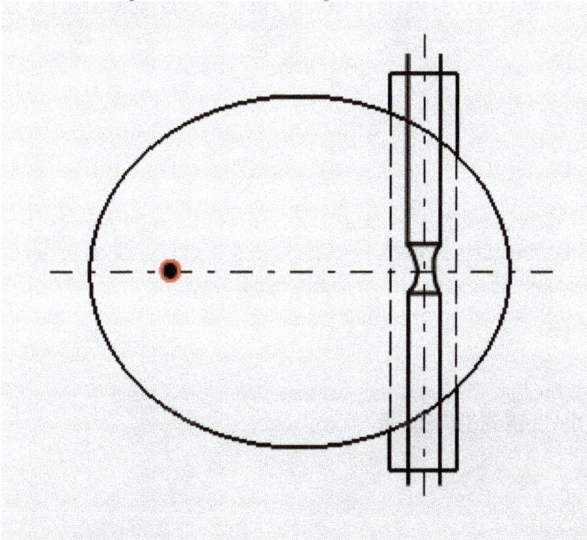

*Fig. 1. Crystal growth apparatus based on a mono-ellipsoidal reflector. The source of light (halogen lamp or xenon lamp) is positioned at the focal point at the left-hand side and is targeting at the melting zone on the right-hand side.*

Illumination configurations for infrared heating based on single ellipsoidal reflectors are used to grow single crystals of oxides, for example [1].

An ellipsoidal mirror serves to concentrate the light of the halogen lamp in the melting zone. Figures 2 and 3 provide a top view and a side view of the reflector, respectively.

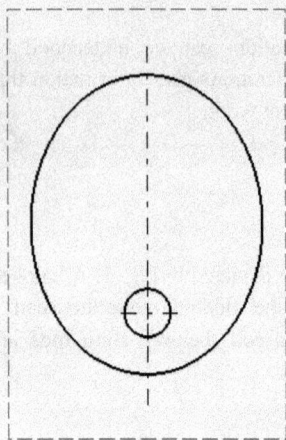

*Fig. 2. Top view of the mirror reflector*

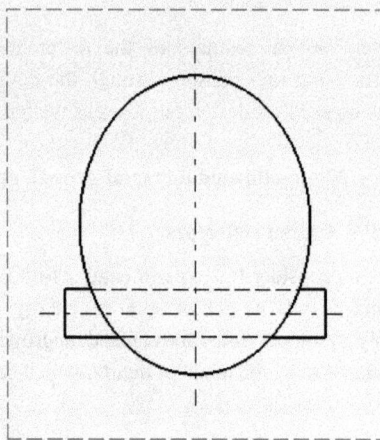

*Fig. 3. Side view of the ellipsoidal and the quartz tube*

During the growth process, the two rods can be rotated along the common axis. Rotating them in opposing directions would enhance mixing while the floating zone remains more stable if the direction of rotation is the same. In any case, rotation of the melting zone has the advantage of exposing changing parts of the material to the inhomogeneous illumination.

In Fig. 4, some rays are depicted that emanate from the source of light. Only one plane that intersects the ellipsoidal reflector is considered.

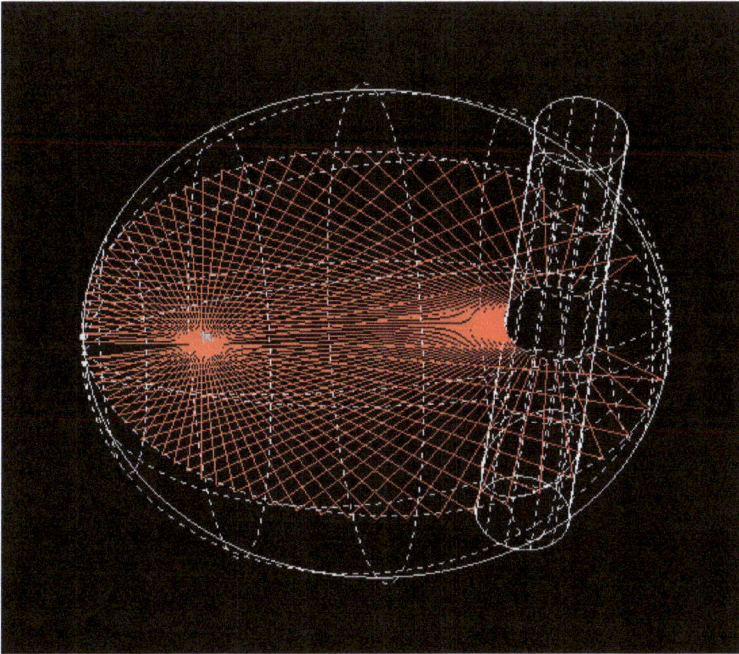

*Fig. 4. Light rays in a plane that is perpendicular to the quartz tube. Only rays starting from the focal point at the left-hand side are considered.*

For clarity, these rays are also shown without reflector contour in Fig. 5. In both figures, the shadowing effect of the absorbing material is obvious. This gives rise to an inhomogeneity in intensity if the rods inside the tube and therefore the melting zone were not rotated.

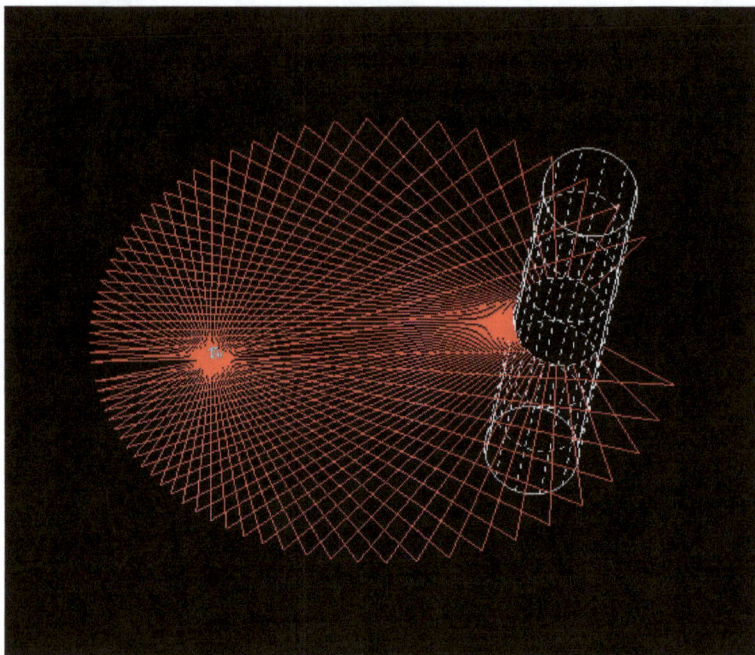

*Fig. 5.    Light pattern formed by rays that emanate from the point-like light source and travel in a plane that is perpendicular to the quartz tube.*

In the following perspective view, some rays in a perpendicular plane are also depicted. It can be seen from the drawing that they are absorbed completely. Some of them reach the rotating rods or the melting zone without being reflected by the ellipsoidal mirror.

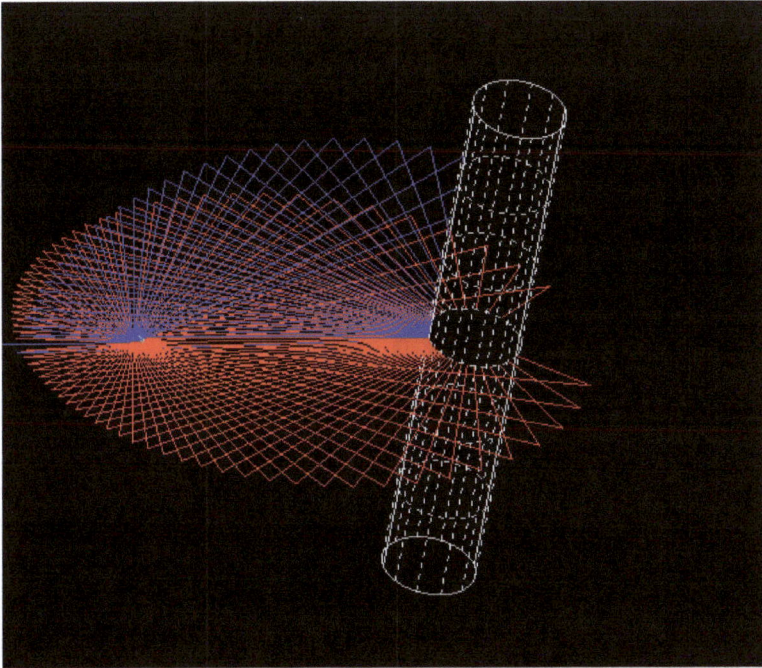

*Fig. 6. Rays starting in the common plane of focal point and glass tube.*

Here, the source of light is sketched as a point source. To get a qualitative impression of an *extended light source* as a halogen lamp or a xenon lamp, for example, it is instructive to have a look at the rays emanated from point sources that are translated with respect to the focal point. An extended source of light might be thought of as a superposition of many of these point-like light sources. And such considerations give a hint where to position an extended source with respect to the ideal focal point in experimental practice.

In the sequence of Fig. 7 to Fig. 12, the position of the light source is therefore varied from a position near the apex of the ellipsoid, through a position at the focal point to a position that is closer to the center of the ellipsoid. This might be considered as a way of scanning a light source that is extended along the optical axis.

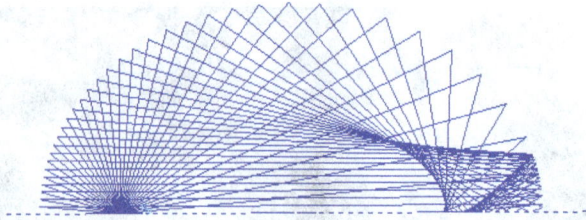

*Fig. 7.     Rays starting at a point that is -6mm away from the focal point.*

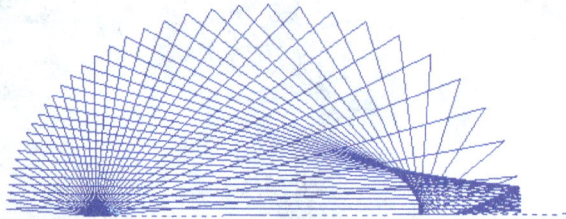

*Fig. 8.     Rays starting at a point that is -4mm away from the focal point.*

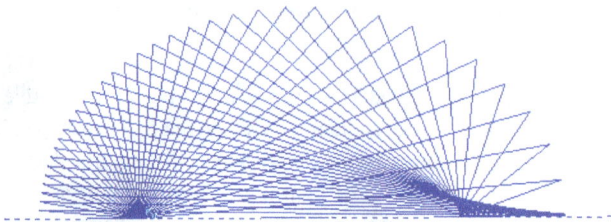

*Fig. 9.     Rays starting at a point that is -2mm away from the focal point.*

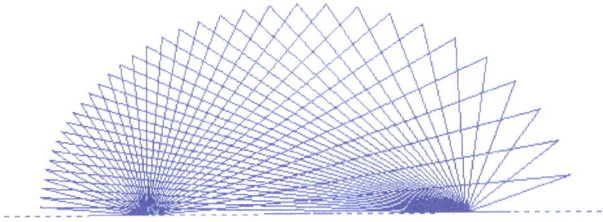

*Fig. 10.    Rays starting at the focal point.*

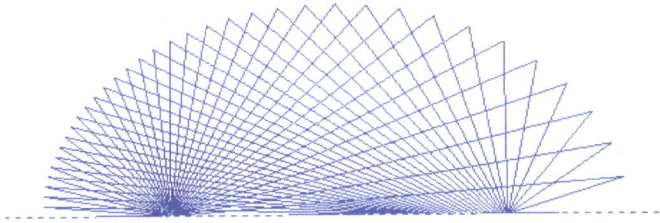

*Fig. 11.    Rays starting at a point that is -+2mm away from the focal point.*

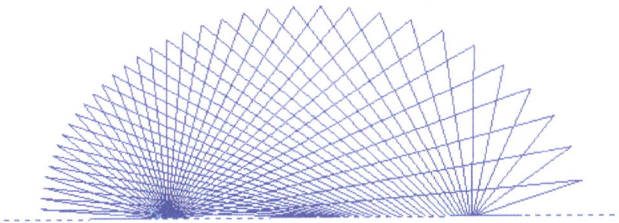

*Fig. 12.    Rays starting at a point that is -+4mm away from the focal point.*

Looking at the Figures 9 to 12, a difference is visible in the ray pattern if the point-like source is translated in the direction of the apex or if it is moved in the direction of the center of the ellipsoid: In the first case, a pronounced caustic is formed while in the second case there is only a slight deviation from the target point.

To gain a better understanding of where the different contributions stem from, it is helpful to distinguish different zones on the optical reflector (Figures 13 to 16). Rays that stem from the region of the mirror that is close to the apex are indicated by red color.

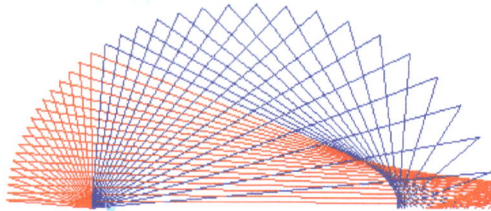

*Fig. 13.    Rays starting at a point that is -4mm away from the focal point.*

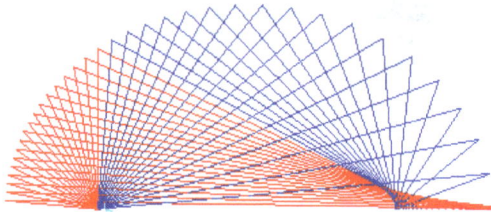

*Fig. 14.    Rays starting at a point that is -2mm away from the focal point.*

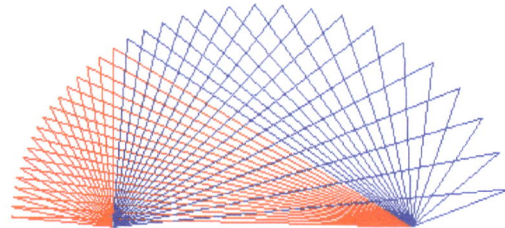

*Fig. 15.    Rays starting at a point that is +2mm away from the focal point.*

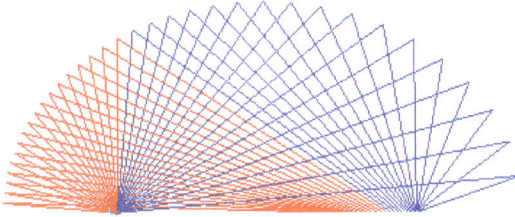

*Fig. 16.    Rays starting at a point that is +4mm away from the focal point.*

It can be concluded from Figures 13 to 16 that rays that emanate from the part of the extended light source that is left of the focal point can be used to illuminate the back part of the sample. This effect is supported by the region of the mirror that surrounds its apex.

In experimental practice, a hole has often to be cut into this region of the mirror to be able to mount the light source. This has as consequence that less rays from this region of the reflector can reach the back part of the melt region.

The sequence of figures also elucidates, why optimizing the position of an extended light source along the optical axis heuristically can have a significant effect on the illumination of the floating zone.

The model of the virtual source helps to better understand why ellipsoidal mirrors can provide an *illumination* in the melt region that is relatively *homogeneous*. The model or geometrical construction is based on the property of an ellipse that the length of each light path that starts at one focal point of the ellipse and ends at the other focal point after a reflection at the ellipse is constant. This implies that the virtual light sources, which correspond to the same focal point but different angles, lay on a circle around the target focal point (Fig. 17). This corresponds to the situation how a point-like light source "is seen" from a point inside the floating zone.

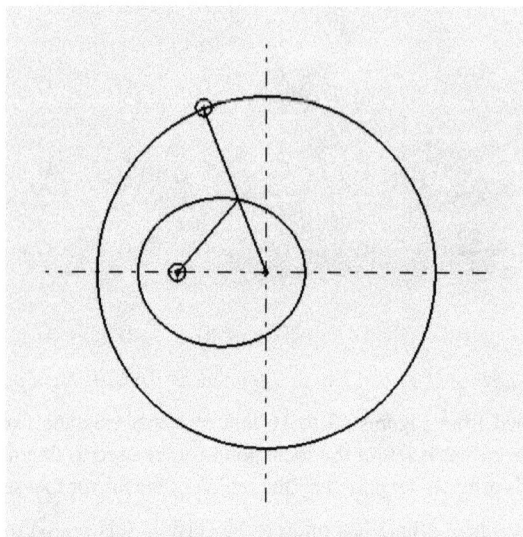

Fig. 17.    Geometrical construction of a virtual point-like light source.

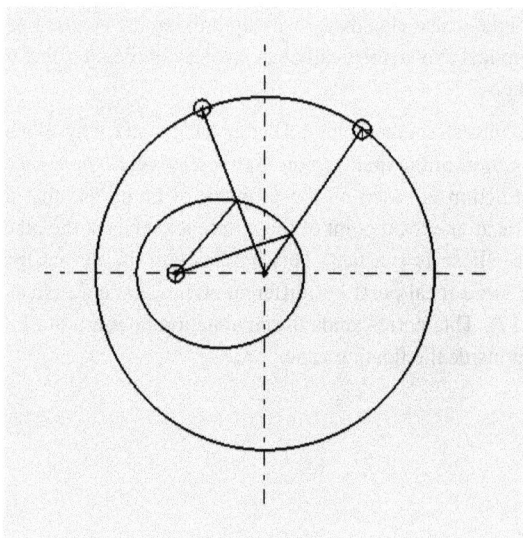

Fig. 18.    Geometrical construction of virtual point-like light sources.

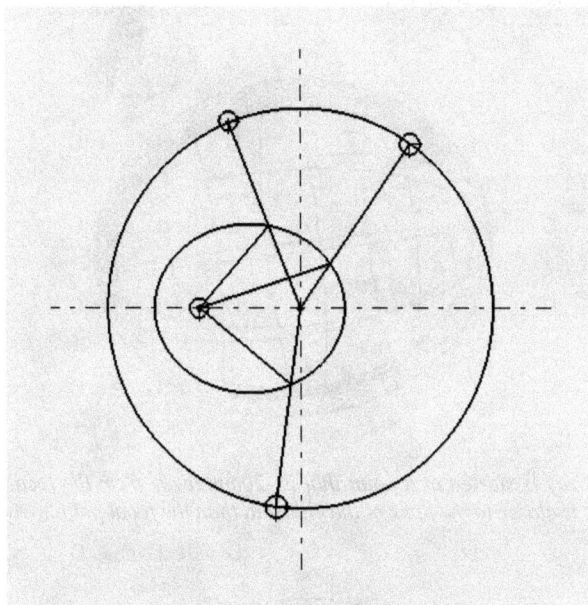

*Fig. 19.    Geometrical construction of virtual point-like light sources.*

Because the virtual light sources are equally distributed around the target, we can conclude that it is homogeneously illuminated in this idealized model.

If the shadowing effect of the rods and of the melting zone itself and the influence of direct light were not taken into account this would correspond to a homogeneous illumination of the melting zone.

In simple cases, it is possible to make "mental ray-tracing", i.e. to predict the path of a ray without starting a numerical simulation, but by making recurrence to optical properties of the corresponding reflectors. Such considerations are useful in the pre-design phase and while doing adjustment work.

Let us have a look at Fig. 20 and Fig. 21:

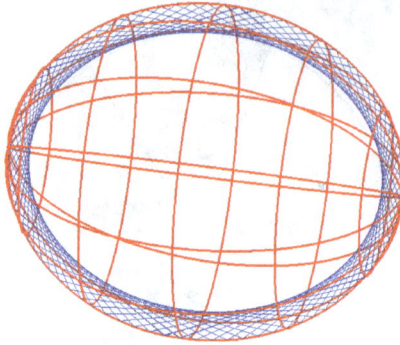

*Fig. 20.     A ray is started at a point that is -20mm away from the focal point, i.e. the source of light is closer to the apex of the ellipsoid than the focal point.*

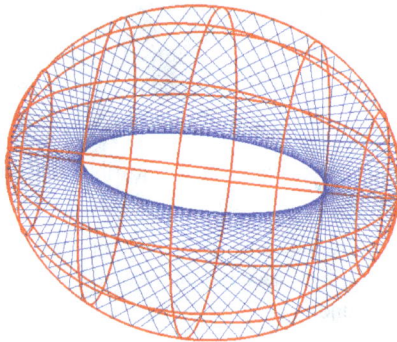

*Fig. 21.     A ray is started at a point that is -2mm away from the focal point, i.e. the source of light is closer to the apex of the ellipsoid than the focal point.*

The general rule can be derived that a ray that starts on the axis and in the region between the apex and the first focal point and that is tangent to an ellipse will remain tangent to this ellipse during its further path.

Fig. 22 and Fig. 23 are given as counterexamples to illustrate that this rule does not hold for rays that start on the axis in the region between the two focal points.

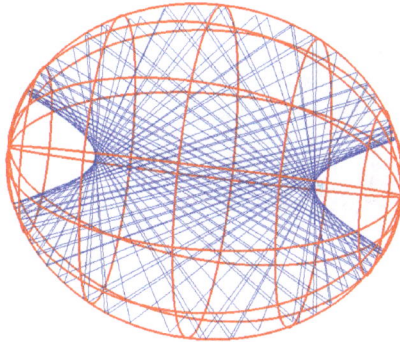

*Fig. 22.    A ray is started at a point that is +2mm away from the focal point, i.e. the source of light is closer to the center of the ellipsoid than the focal point.*

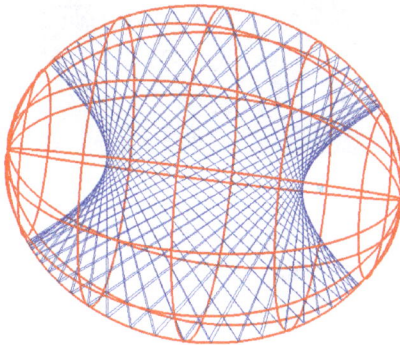

*Fig. 23.    A ray is started at a point that is +6mm away from the focal point, i.e. the source of light is closer to the center of the ellipsoid than the focal point.*

## 2.2    Open reflectors

In crystal growth experiments, mono-ellipsoids that are closed (with the exception of the aperture for the tube) are often used, but we will also consider open reflectors of this type for completeness.

In Fig. 24, a reflector is depicted that is made up of two parts of an ellipsoid. The light emanating from the source of light is absorbed by the rods and the floating zone inside the quartz tube.

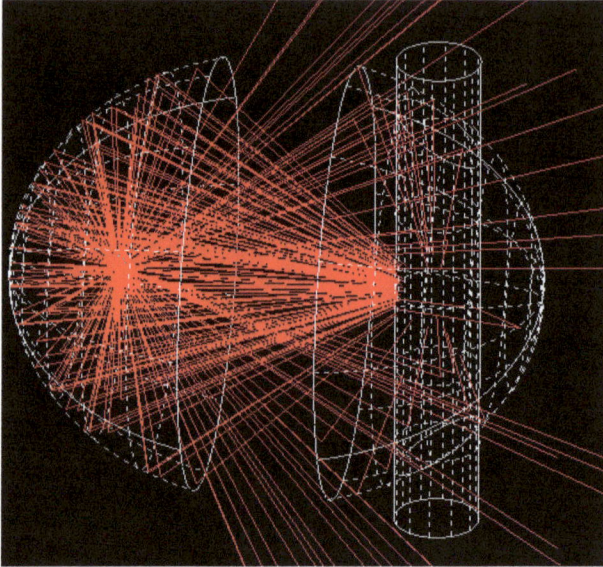

*Fig. 24.    An open ellipsoidal reflector that serves to illuminate the floating zone inside a quartz tube. The light rays emanating from the focal point of the ellipsoid are generated in a random way.*

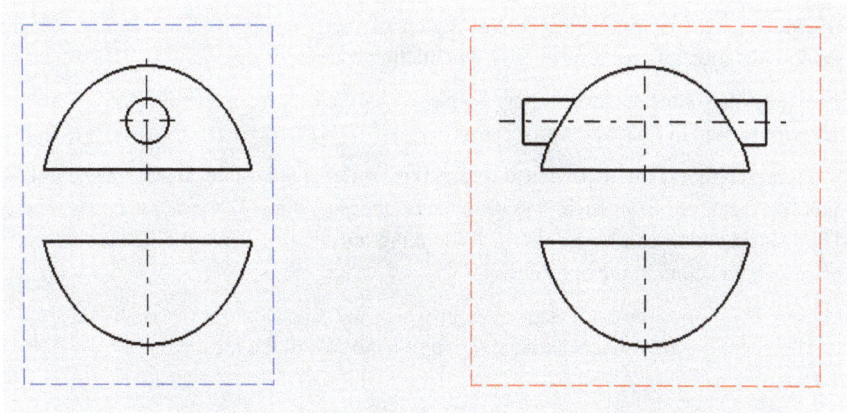

*Fig. 25 and 26.    Two views on an open ellipsoid reflector (top view and side view).*

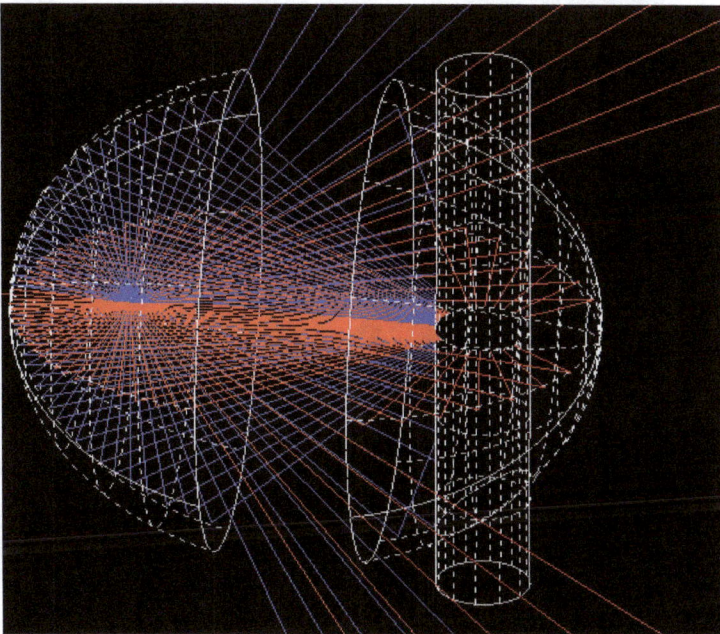

*Fig. 27.    A partial ellipsoid mirror illuminating a specimen inside a quartz tube. The light rays that travel in two perpendicular planes are shown.*

Obviously, light is lost in the central region of the two mirrors. And in Fig. 27, the shadowing effect of quartz tube, rods, and melting zone is visible.

The following considerations apply to closed as well as to open reflectors. A partial reflector is used in illustrations for clarity.

In Figures 9 to 12, the dependence on the position of a point-like source on the optical axis was displayed in an attempt to get a better understanding of an *extended light source*. This consideration can be continued by looking at rays that leave the light source in a plane perpendicular to the optical axis.

Figures 28 to 30 represent a sequence, in which the point-like source is shifted from a position near the apex to a position at the right-hand side of the focal point.

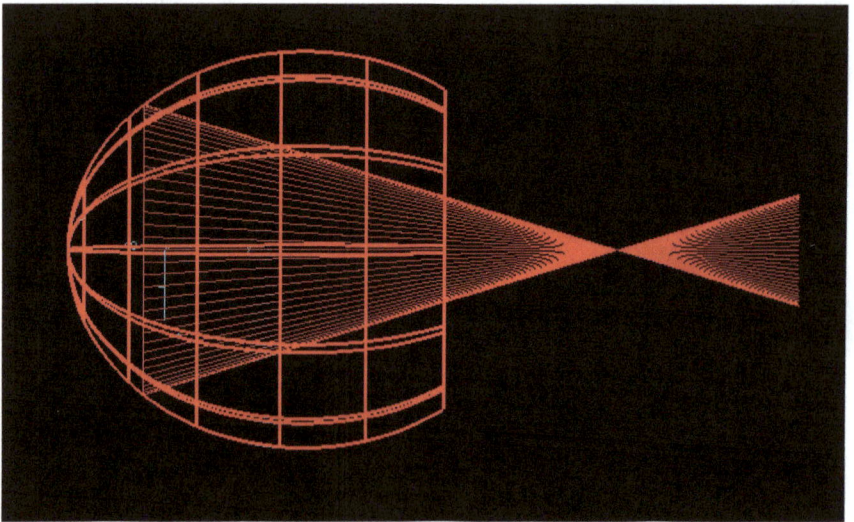

*Fig. 28. Rays that start from a point that is translated by -6mm with respect to the focal point and in a plane that is perpendicular to the optical axis of the ellipsoidal mirror.*

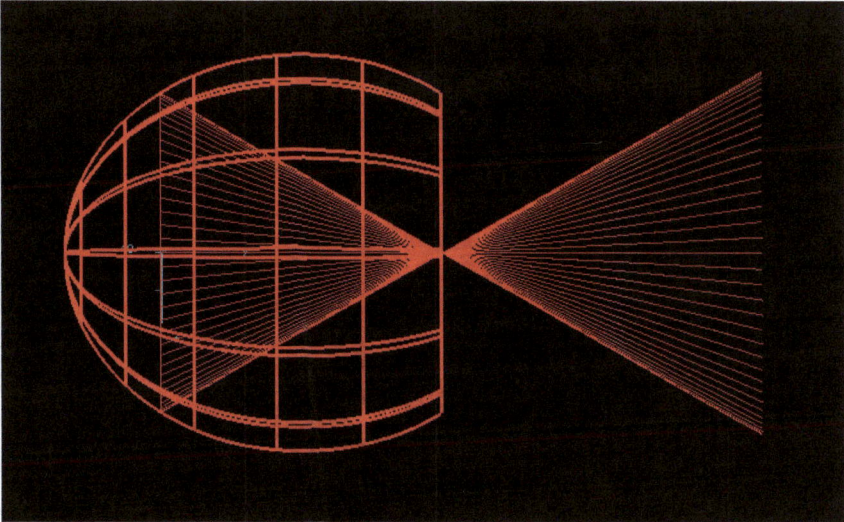

*Fig. 29.    Rays that start from the focal point and in a plane that is perpendicular to the optical axis of the ellipsoidal mirror.*

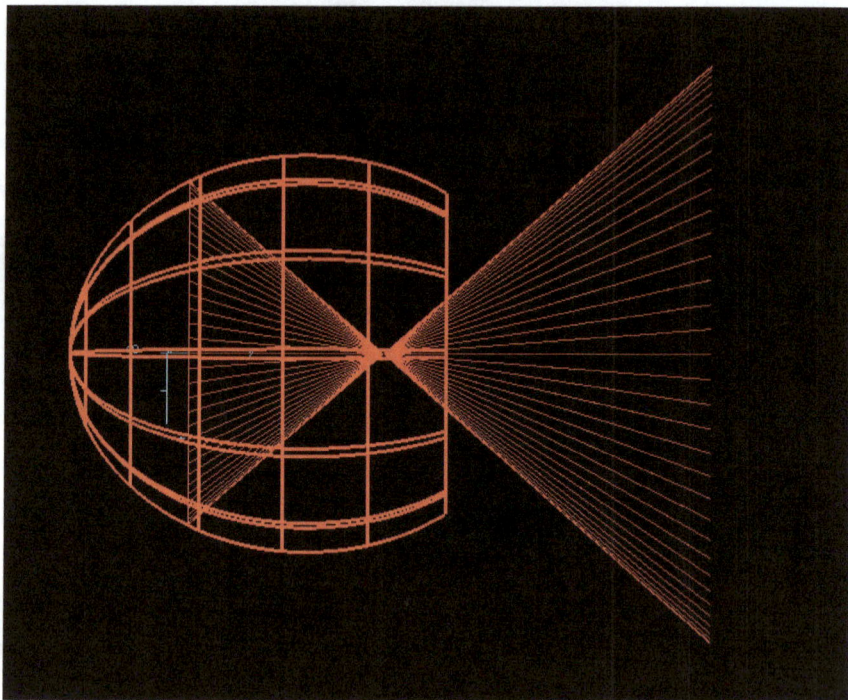

*Fig. 30.        Rays that start from a point that is translated by +6mm with respect to the focal point and in a plane that is perpendicular to the optical axis of the ellipsoidal mirror.*

The sequence shows that in all the three cases the rays, which stem from a circle on the surface of the ellipsoid, intersect the optic axis at a common point. The position of this point varies according to the position of the point-like light-source on the optic axis.

The situation gets more complicated if the source of light is also translated laterally with respect to the optic axis and then shifted (Figs. 34-36).

First, we have a look at the simpler situation that the light source is only translated laterally with respect to the focal point (Figs. 31-33).

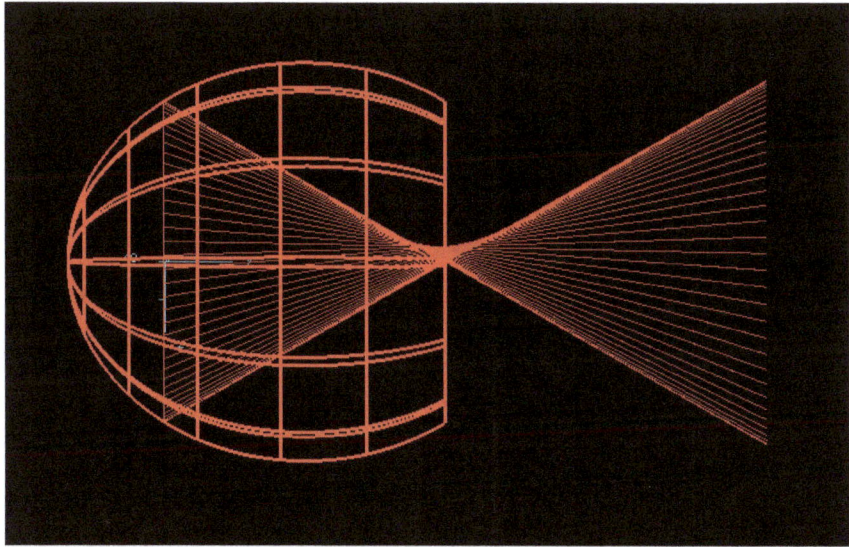

*Fig. 31.* *Rays that start from a point that is translated <u>laterally</u> by +2mm with respect to the focal point and in a plane that is perpendicular to the optical axis of the ellipsoidal mirror.*

The rays, which stem from a circle on the ellipsoid, do no longer intersect at a common point, the focus. Instead, a caustic surface is formed in the focal region.

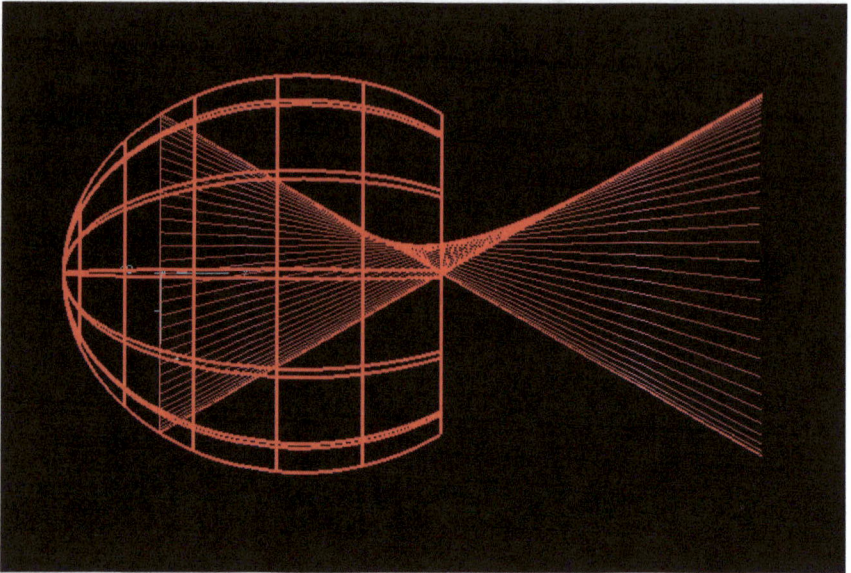

*Fig. 32.* *Rays that start from a point that is translated <u>laterally</u> by +4mm with respect to the focal point and in a plane that is perpendicular to the optical axis of the ellipsoidal mirror.*

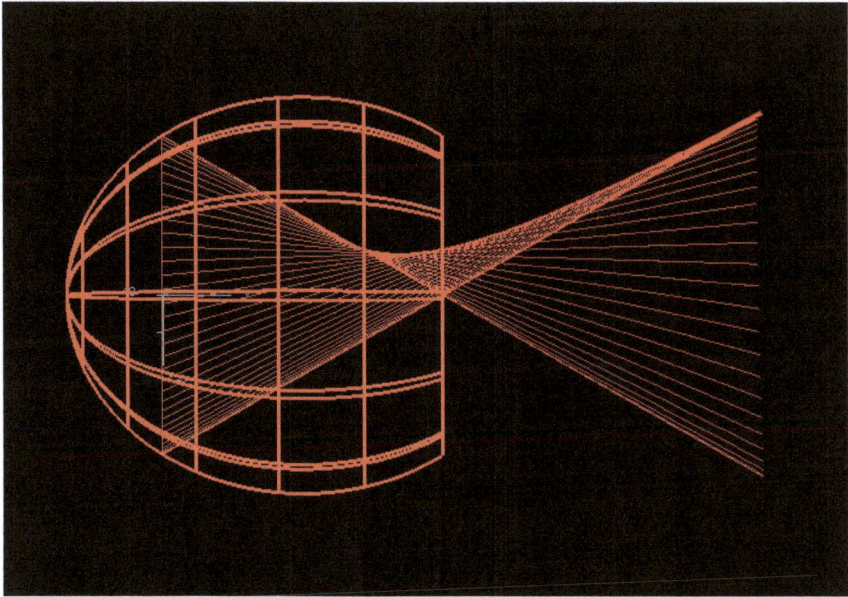

*Fig. 33.* *Rays that start from a point that is translated <u>laterally</u> by +6mm with respect to the focal point and in a plane that is perpendicular to the optical axis of the ellipsoidal mirror.*

To provide a three-dimensional impression of the corresponding caustics, they are also shown in a perspective view in Figures 34 to 36. This is one of these situations where it is easier to describe what happens using pictures than using words or formulae.

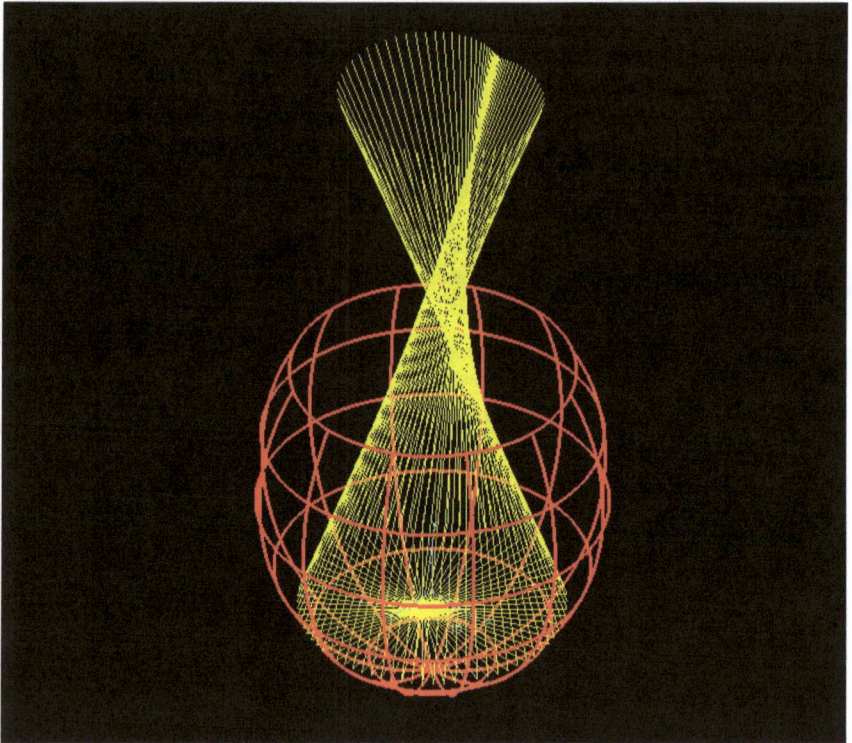

*Fig. 34.* *Perspective view of rays that start from a point that is translated by -4mm along the optical axis and <u>laterally</u> by +4mm with respect to the focal point. The rays start in a plane that is perpendicular to the optical axis of the ellipsoidal mirror.*

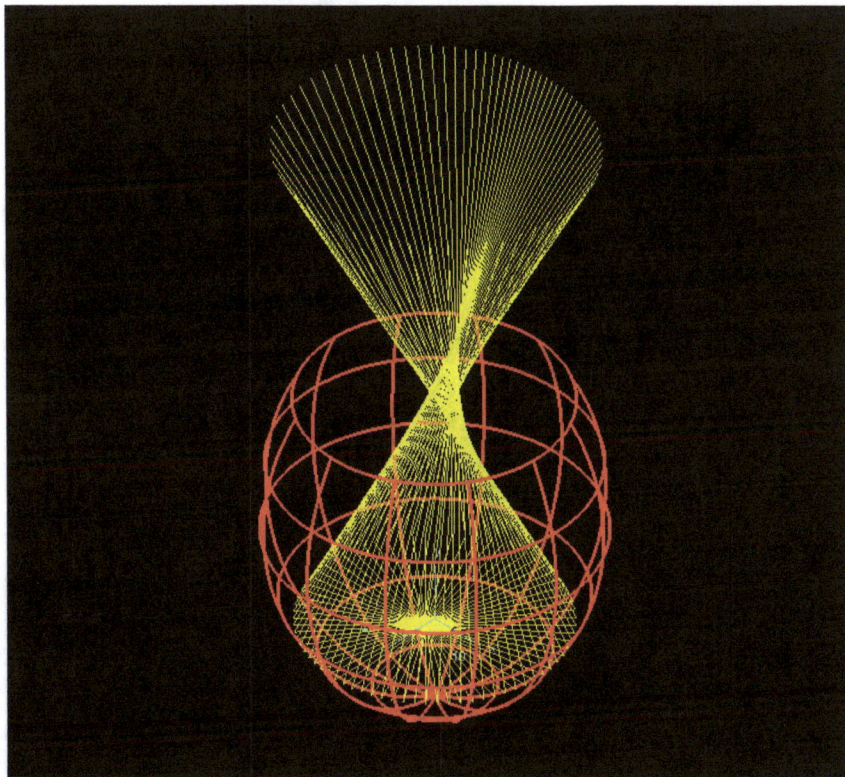

*Fig. 35.    Perspective view of rays that start from a point that is translated <u>laterally</u> by +4mm with respect to the focal point. The rays start in a plane that is perpendicular to the optical axis of the ellipsoidal mirror*

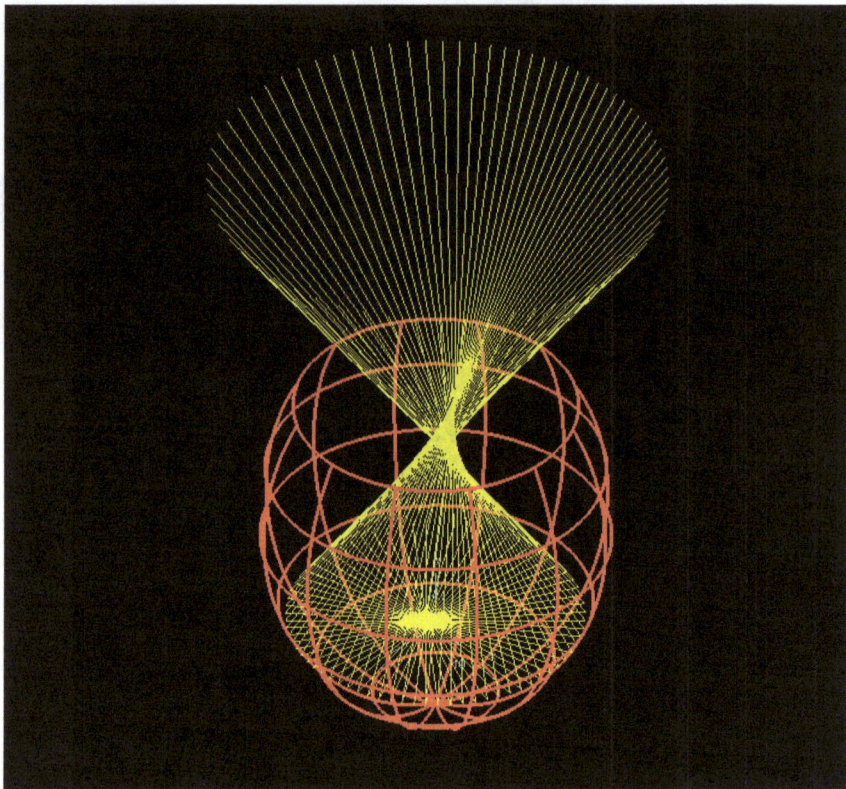

*Fig. 36.  Perspective view of rays that start from a point that is translated by +4mm along the optical axis and <u>laterally</u> by +4mm with respect to the focal point. The rays start in a plane that is perpendicular to the optical axis of the ellipsoidal mirror.*

The effect of these displacement can alternatively be visualized by looking at the light distribution in the focal plane. This gives an indication on how the floating zone is being illuminated. In other optical experiments, it would be possible to hold a piece of paper in a plane perpendicular to the optical axis and this is frequently done to align low-power lasers, for example. In the set-up for a crystal-growth experiment it is not possible, because the light sources are so intense.

Figures 37 to 40 are a sequence of light distributions on a virtual screen in the second focal plane that correspond to focusing a point-like source of light.

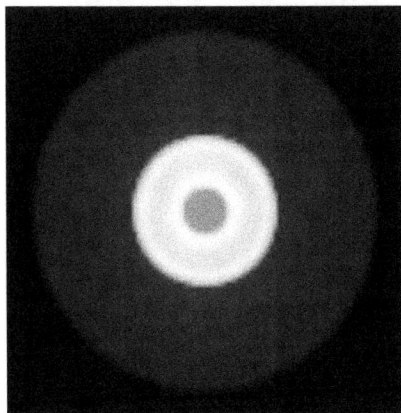

*Fig. 37.    Light distribution in the second focal plane when a point-like light source is translated by -6mm along the optical axis with respect to the focal point.*

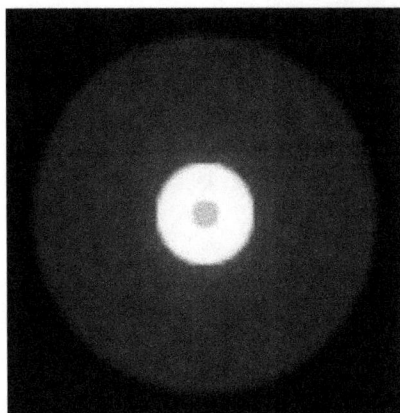

*Fig. 38.    Light distribution in the second focal plane when a point-like light source is translated by -4mm along the optical axis with respect to the focal point.*

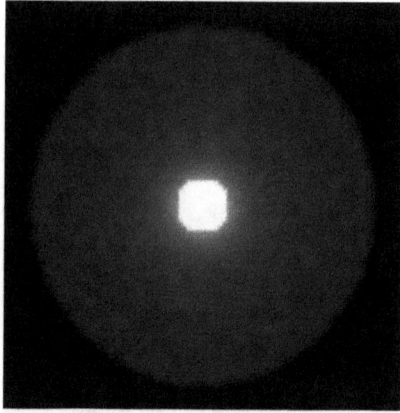

*Fig. 39.* *Light distribution in the second focal plane when a point-like light source is translated by -2mm along the optical axis with respect to the focal point.*

*Fig. 40.* *Light distribution in the second focal plane when a point-like light source is at the focal point.*

Figures 41 to 44 are intended to illustrate the effect of focusing a point-like light-source that is off-axis, i.e. translated laterally with respect to the optic axis.

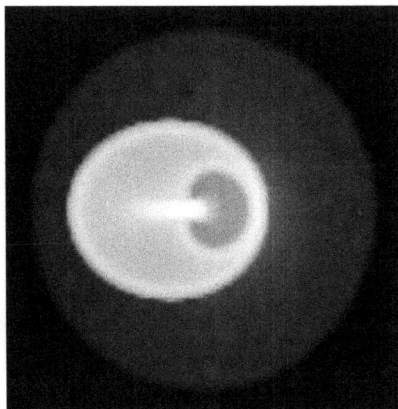

*Fig. 41.     Light distribution in the second focal plane when a point-like light source is translated by -6mm along the optical axis and by 6mm laterally with respect to the focal point.*

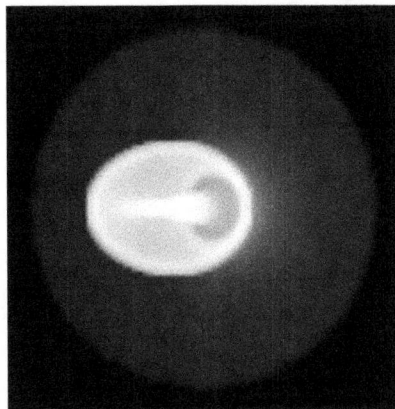

*Fig. 42.     Light distribution in the second focal plane when a point-like light source is translated by -4mm along the optical axis and by 6mm laterally with respect to the focal point.*

*Fig. 43.* *Light distribution in the second focal plane when a point-like light source is translated by -2mm along the optical axis and by 6mm laterally with respect to the focal point.*

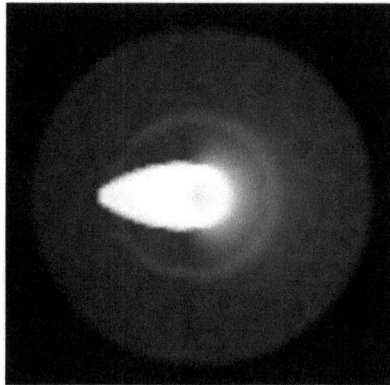

*Fig. 44.* *Light distribution in the second focal plane when a point-like light source is translated by 6mm laterally with respect to the focal point.*

Real light distributions are a superposition of all these individual contributions.

We therefore turn to a way that allows for a more summary representation of optical ray paths.

*Caustics* are a fascinating tool to characterize optical phenomena and monographs are devoted to their study [3]. Here, simple caustics are contemplated and compared to get an impression of the light distribution generated by the ellipsoidal mirror.

As before, a distinction has to be made between direct light and light that is reflected by the surface of the ellipsoidal mirror (Fig. 45).

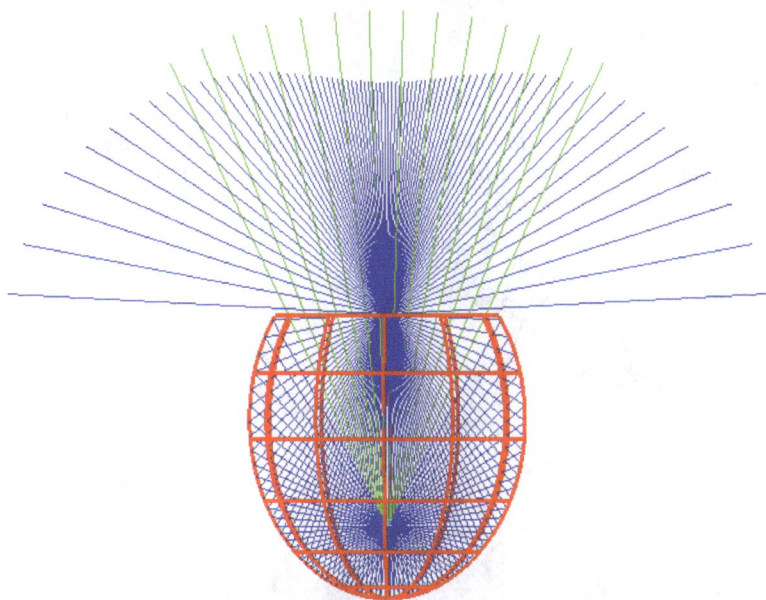

*Fig. 45.     Caustic of a point-like light source at the focal point. Direct light, which is not reflected by the mirror, is indicated by green color.*

Looking at the sequence of Fig. 46 to Fig. 51, we can see how this caustic changes when the point-like source of light is shifted along the optic axis of the mirror.

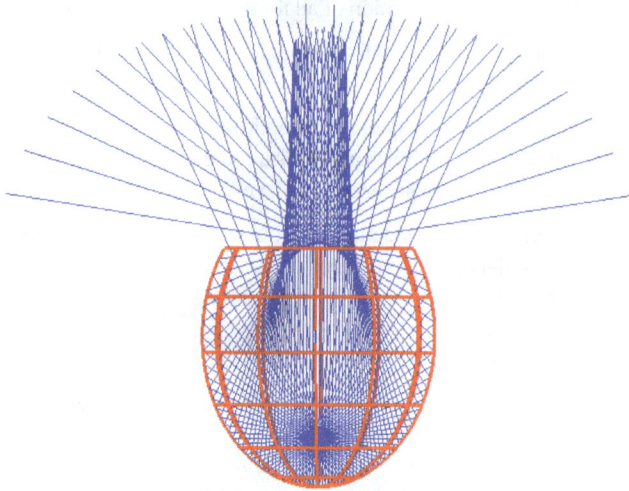

*Fig. 46.     Caustic of a point-like light source that is translated by -6mm with respect to the focal point along the optical axis. This corresponds to a movement in the direction of the apex of the ellipsoidal reflector.*

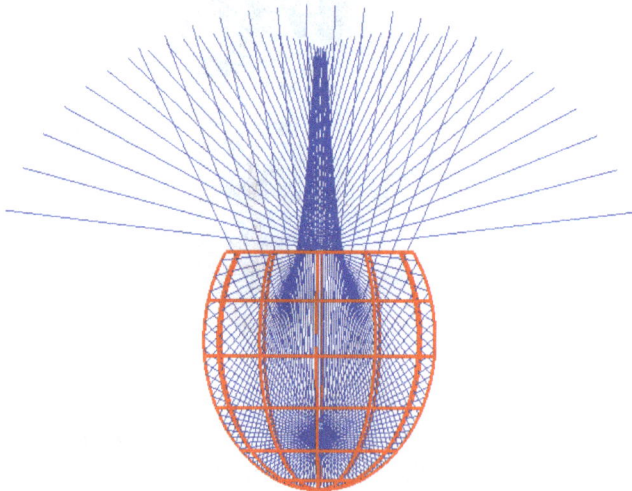

*Fig. 47.     Caustic of a point-like light source that is translated by -4mm with respect to the focal point along the optical axis. This corresponds to a movement in the direction of the apex of the ellipsoidal reflector.*

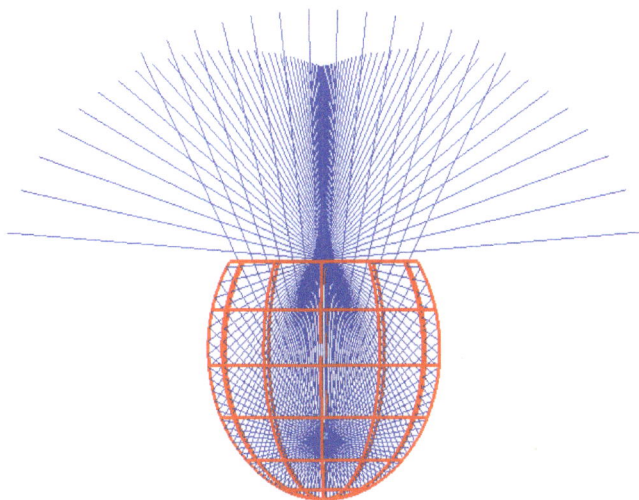

Fig. 48.    Caustic of a point-like light source that is translated by -2mm with respect to the focal point along the optical axis. This corresponds to a movement in the direction of the apex of the ellipsoidal reflector.

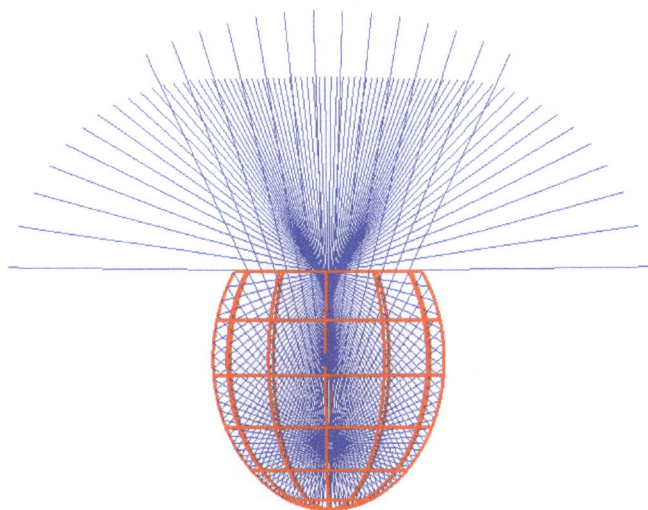

Fig. 49.    Caustic of a point-like light source that is translated by +2mm with respect to the focal point along the optical axis. This corresponds to a movement in the direction of the center of the ellipsoidal reflector.

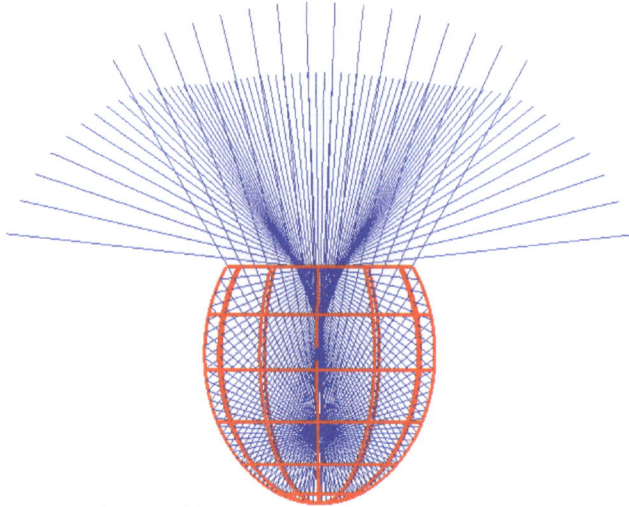

*Fig. 50.* *Caustic of a point-like light source that is translated by +4mm with respect to the focal point along the optical axis. This corresponds to a movement in the direction of the center of the ellipsoidal reflector.*

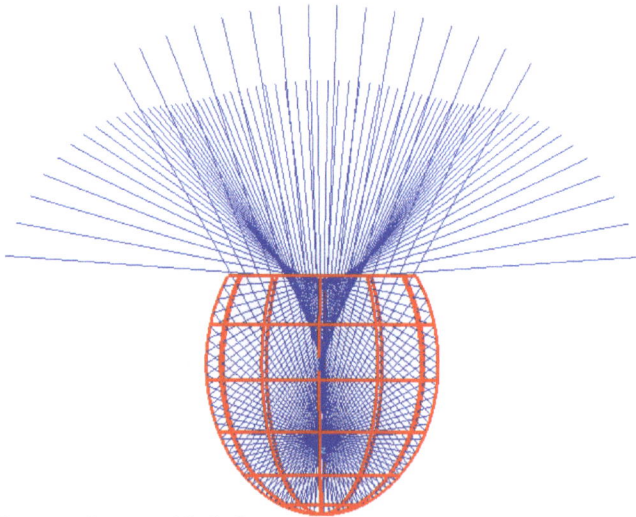

*Fig. 51.* *Caustic of a point-like light source that is translated by +6mm with respect to the focal point along the optical axis. This corresponds to a movement in the direction of the center of the ellipsoidal reflector.*

In Fig. 52 and Fig. 53, we turn to a more complicated situation, namely to the consideration of a point-like light-source that is also laterally translated with respect to the axis of symmetry.

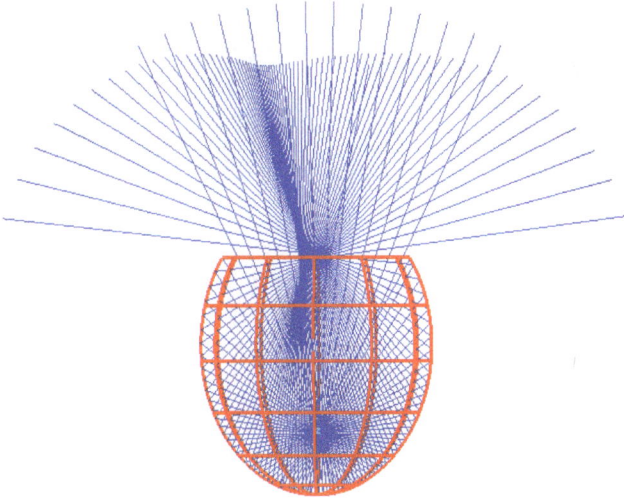

*Fig. 52.    Caustic of a point-like light source that is translated by +2mm laterally with respect to the focal point.*

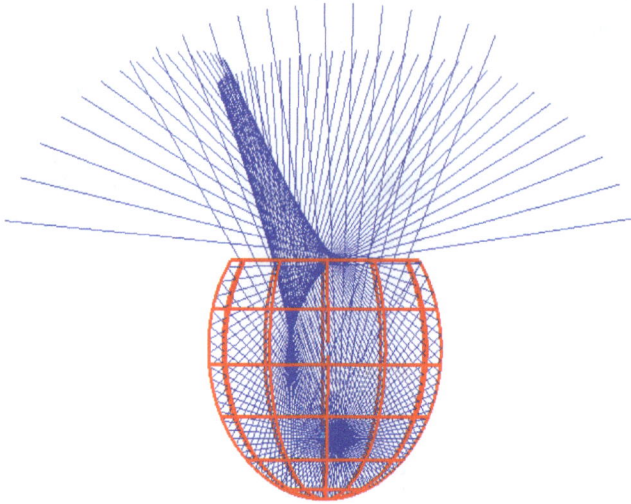

*Fig. 53.    Caustic of a point-like light source that is translated by +6mm <u>laterally</u> with respect to the focal point.*

In these caustics, the vertical line is no longer an axis of symmetry. This can also be seen from the corresponding three-dimensional representations (Figs. 54-56).

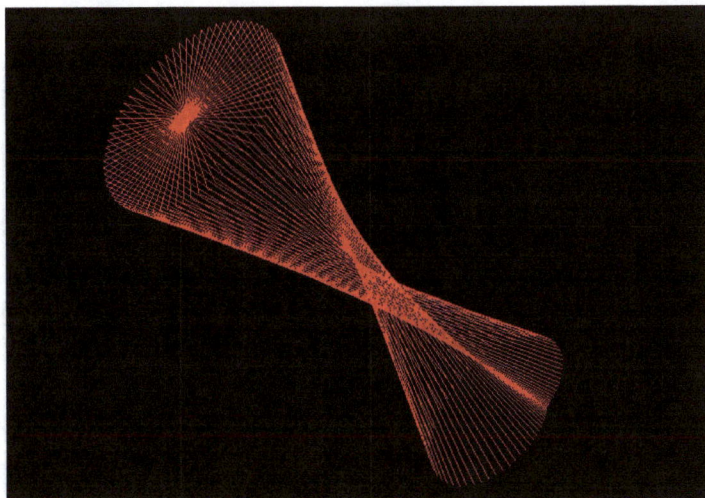

*Fig. 54.* *Caustic of a point-like light source that is translated by -4mm along the optical axis and by +4mm laterally with respect to the focal point.*

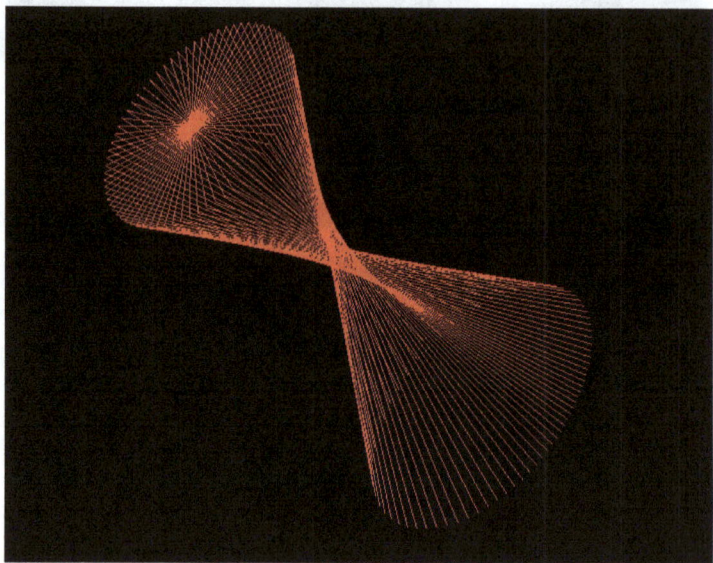

*Fig. 55.* *Caustic of a point-like source that is translated laterally by +4mm with respect to the focal point.*

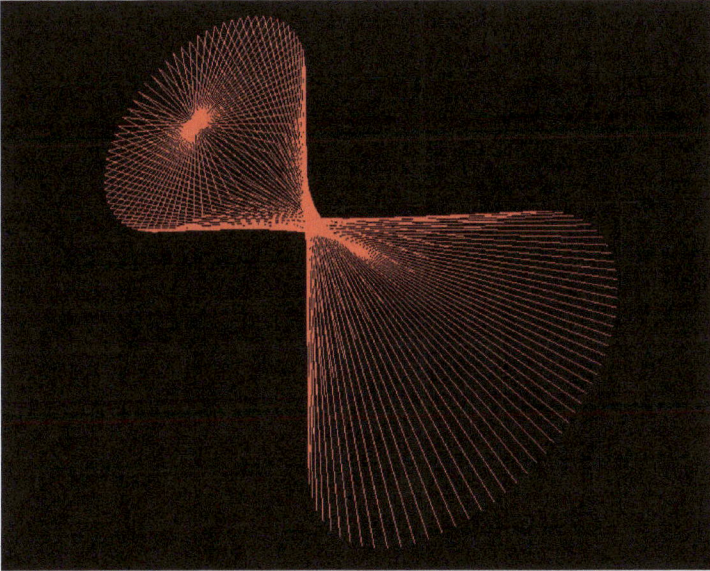

*Fig. 56.      Caustic of a point-like light source that is translated by +4mm along the optical axis and by +4mm laterally with respect to the focal point.*

They are collected in Fig. 57 and indicated by different colors.

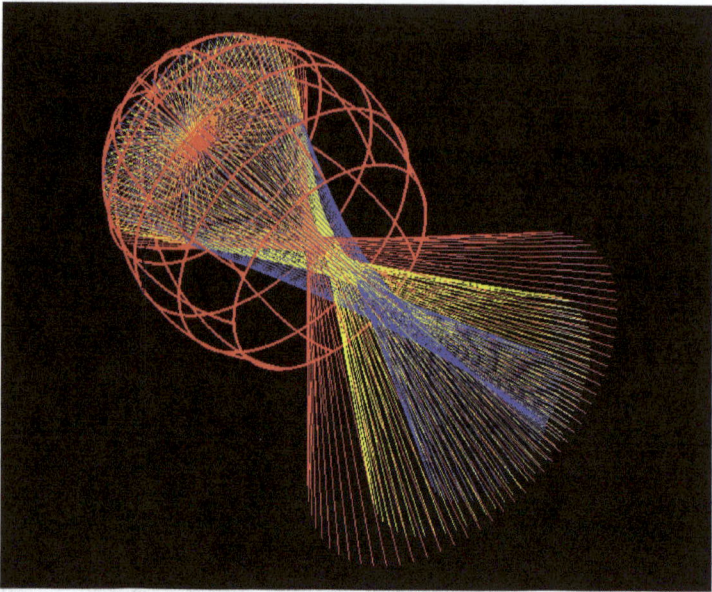

*Fig. 57.*     *Caustics of point-like light sources that are translated by +4mm laterally with respect to the focal point and by -4mm, 0mm, and +4mm, respectively, along the optical axis.*

### 3. Crystal-growth apparatus using multiple ellipsoids

Imaging furnaces can be built using single ellipsoidal mirrors or double mirror systems. In addition, experimental configurations with more than two ellipsoidal reflectors are in use. Double mirror systems for optical heating can be based on ellipsoidal mirrors as well as on paraboloidal reflectors [4].

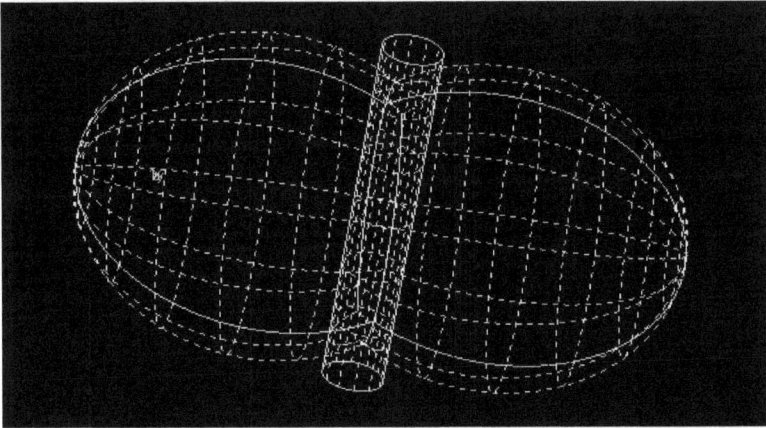

*Fig. 58.    Double ellipsoidal arrangement: Two partial ellipsoid mirrors illuminate a floating zone at the center of symmetry.*

Double-ellipsoid reflectors are often applied in zone melting for crystal growth [1,5-8].

They allow for a more symmetric illumination of the floating zone inside the tube made of quartz (Fig. 59).

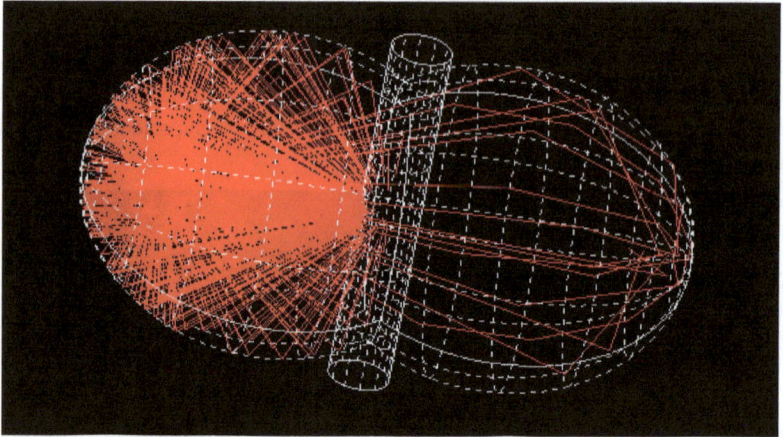

*Fig. 59.     Double ellipsoidal arrangement: Light rays are shown that come from the focal point at the left-hand side. During crystal growth, both light sources are switched on and illuminate the floating zone with intense light.*

A further step to increase the symmetry of the illumination of the floating zone consists of using *four reflectors* (Fig. 60). This measure obviously increases the homogeneity of the intensity distribution in the floating zone.

A disadvantage might be seen in the increased complexity of the optical arrangement.

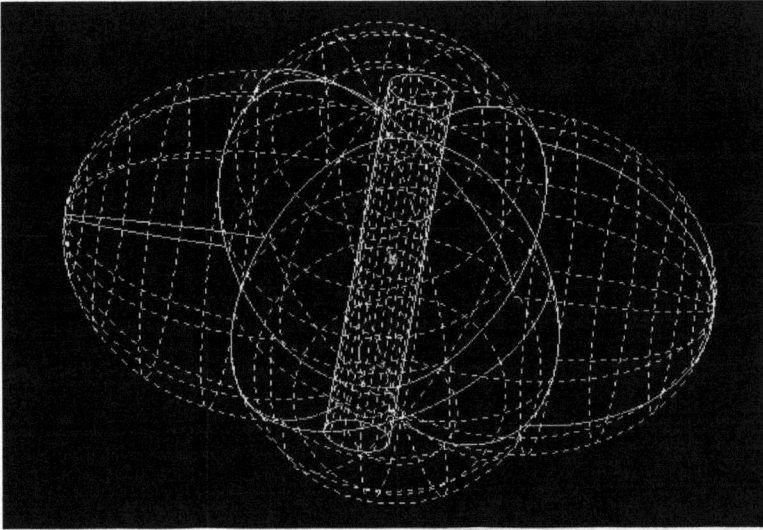

*Fig. 60.* *Configuration based on four partial ellipsoidal reflectors.*

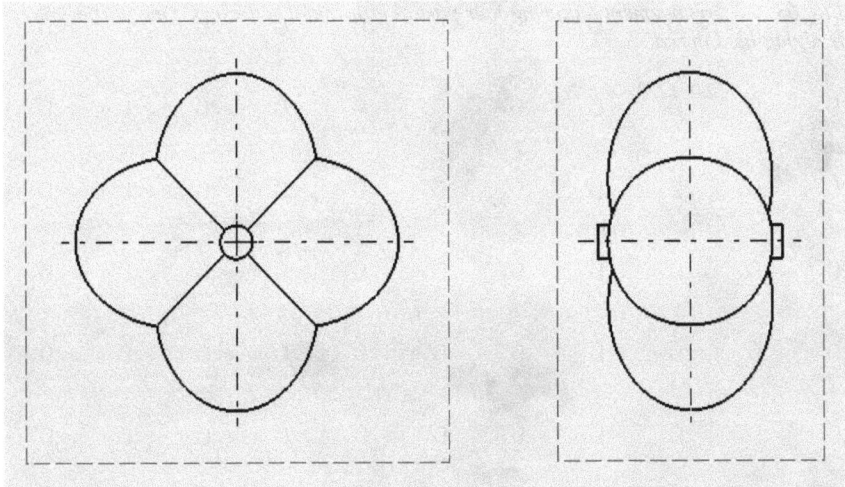

*Figures 61 and 62.* *Top view and side view of a configuration based on four partial ellipsoidal reflectors*

In the optical configuration with fourfold symmetry, four partial ellipsoid reflectors are arranged around a common axis through a common focal point. Each of the mirrors is equipped with a light source to heat the sample positioned at the common focal point.

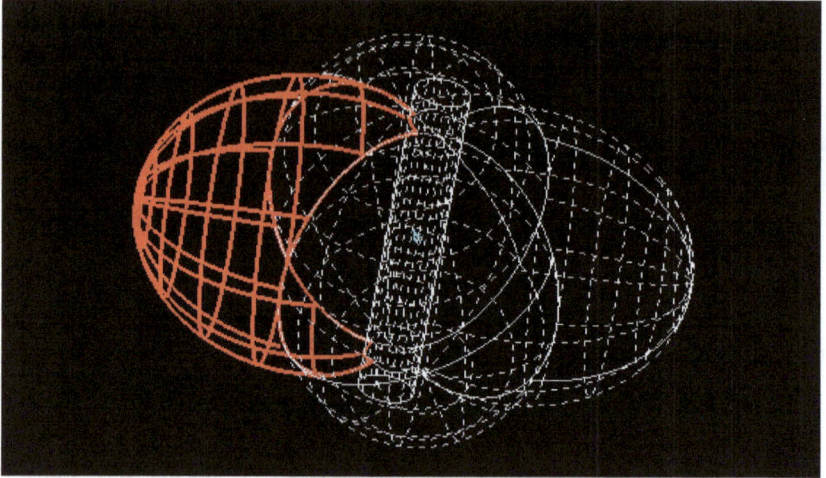

*Fig. 63.     Arrangement featuring four partial ellipsoidal reflectors. One partial mirror is highlighted in red.*

*Fig. 64.     Infrared imaging furnace for crystal growth featuring four reflectors. Image courtesy of Quantum Design, Inc. © 2016.*

Eyer, Nitsche and Zimmermann [10] provide a theoretical comparison of illumination configurations for crystal growth with single and double ellipsoids as well as multi-ellipsoids with three and four chambers.

Finally, we have a look at reflector arrangements with *six-fold symmetry*. The corresponding mirrors are sketched in Figures 65 to 69. To stress the pronounced influence of this shape factor, Figs. 65-66 and Figs. 67-68 have different values of eccentricity.

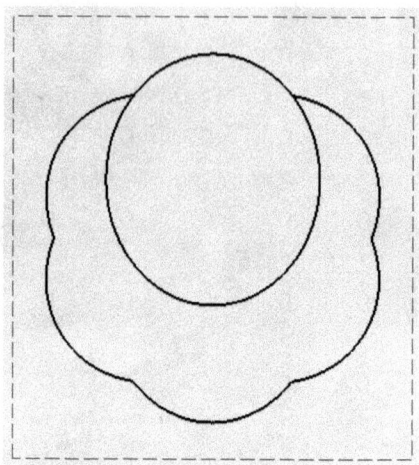

Fig. 65.    Constructing a compound mirror with six-fold symmetry.-

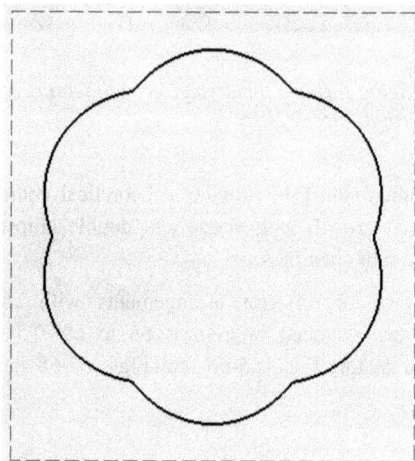

Fig. 66.    Compound mirror with six-fold symmetry.

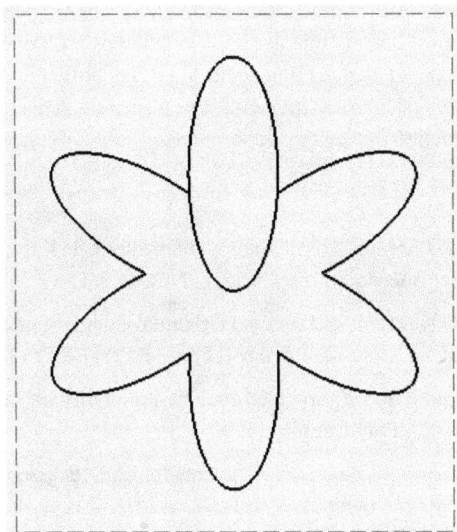

*Fig. 67.    Constructing a compound mirror with six-fold symmetry.*

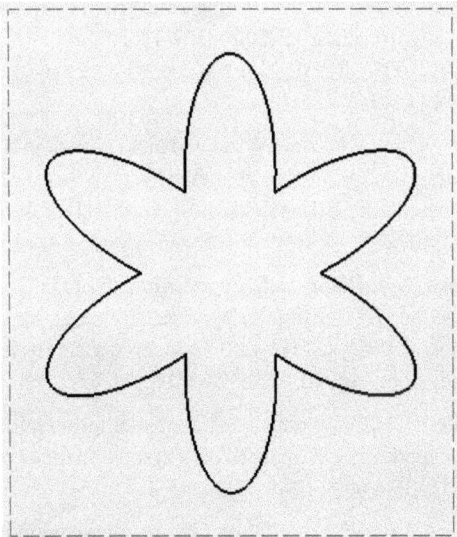

*Fig. 68.    Compound mirror with six-fold symmetry.*

Even though this example might seem academic at first glance, there are indeed configurations for optical heating that use six xenon lamps positioned in regular intervals of 60° around the axis of the melting zone [11]. To further increase the light flux, eighteen additional tungsten halogen lamps are also mounted above and below each xenon lamp and directed to the melting zone.

**References**

[1]     T. Mizutani, K. Matsumi, H. Makino, T. Yamamoto, T. Kato, Single crystal growing apparatus using infrared heating, NEC Res. Dev.. 33 (1974) 88-92.

[2]     K. Kamiryo, T. Kano, H. Matsuzawa, Optimum design of elliptical pumping chambers for solid lasers, Jap- J- Appl. Phys. 6(12) (1966) 1217-1226.

[3]     J-F. Nye, Natural Focusing and the Fine Structure of Light, Institue of Physics Publishing, Bristol, Philadelphia, 1999.

[4]     P.E. Glaser, Imaging-furnace developments for high-temperature research, J. of the Electrochemical Society 107 (1960) 226-231. https://doi.org/10.1149/1.2427656

[5]     C.W. Lan, J.C. Leu, Y. Huang, On the design of double-ellipsoid mirror furnace and its thermal characteristics for floating-zone growth of $Sr_xBa_{1-x}TiO_2$ single crystals, Cryst. Res. Technol. 35 (2000) 167-176. https://doi.org/10.1002/(SICI)1521-4079(200002)35:2<167::AID-CRAT167>3.0.CO;2-9

[6]     A. Miyazaki, H. Kimura, X. Jia, Crystal and glass growth in $BaO-B_2O_3-Al_2O_3$ system by floating zone pulling down method, Cryst. Res. Technol. 35 (2000) 1245-1250. https://doi.org/10.1002/1521-4079(200011)35:11/12<1245::AID-CRAT1245>3.0.CO;2-R

[7]     H.A. Dabkowska, A.B. Dabkowski, Crystal Growth of Oxides by Optical Floating Zone Technique, in: G. Dhanaraj, K. Byrappa, V. Prasad, M. Dudley (Eds.), Springer Handbook of Crystal Growth, Springer-Verlag, Berlin, Heidelberg, 2010, pp. 367-391. https://doi.org/10.1007/978-3-540-74761-1_12

[8]     A. Eyer, R. Nitsche, H. Zimmermann, A double-ellipsoid mirror furnace for zone crystallization experiments in Spacelab, J. Cryst. Growth. 47 (1979) 219-229. https://doi.org/10.1016/0022-0248(79)90245-8

[9]     L. Wondraczek, J. Deubener, U. Lohse, Containerless melting of optical glasses by thermal imaging, Glass Sci. Technol. 78C (2005) 54-59. *6 reflector parts*

# CHAPTER 4

# Imaging furnaces with intermediate focus

## Abstract

In some laboratories, experimental set-ups for crystal growth are in use that feature optical systems with an intermediate focus. This focal point is shared by two neighboring optical reflectors. This experimental approach is different from direct illumination of the sample. One might think of the first ellipsoidal mirror as an optical unit that pre-shapes the light distribution and directs it in a direction along the optical axis. The second ellipsoidal mirror might be interpreted as an optical unit that receives the light and concentrates it in the narrow region where it is used for melting the specimen.

## Keywords

Crystal-growth apparatus, double ellipsoid, intermediate focal point, compound mirror, floating-zone technique.

## Contents

1. Introduction ........................................................................................ 78

2. Horizontal design ............................................................................... 78

2.1 General ............................................................................................... 78

2.2 Dedicated set-up for crystal growth ................................................. 81

3. Vertical design .................................................................................... 84

References ................................................................................................. 87

## 1.    Introduction

There are different groups of experimental configurations that are in use to grow single crystals with the optical floating zone method. In chapter 3, optical arrangements were illustrated that feature direct illumination of the melting zone. In this chapter, devices are presented that are built up in such a way that two optical reflectors have a focal point in common, which might be designated as intermediate focus.

Depending on the orientation of the axes of the two ellipsoid mirrors in space, this group of experimental configurations with intermediate focus can again be sub-divided in horizontal designs and vertical designs.

## 2.    Horizontal design
## 2.1    General

To illustrate the working principle of double ellipsoids with intermediate focus, let us start with an idealized example. Fig. 1 shows a closed reflector that is made up of two symmetric parts. Both parts have a common focal point. This implies that a part of the light rays that starts from one of the outer focal points is first concentrated at an intermediate focus before being concentrated again at the corresponding outer focal point. This does not hold for those rays that leave the first focal point without hitting the first reflector.

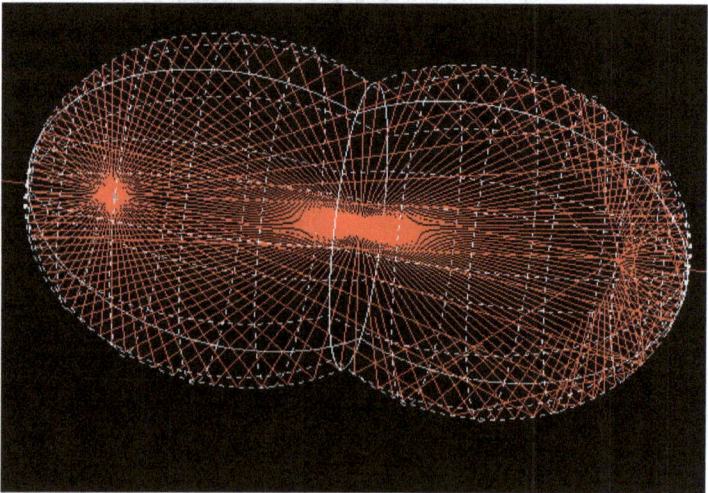

*Fig. 1.    Idealized compound mirror made up of two symmetric ellipsoidal parts. Rays are shown that emanate from a focal point on the left-hand side. A part of these rays is re-collected at the focal point on the right-hand side.*

In Fig. 2, a distinction is made between those rays that impinge on the first ellipsoid mirror after leaving the source of light and those rays that pass through the intermediate focal point without being reflected by the first reflector.

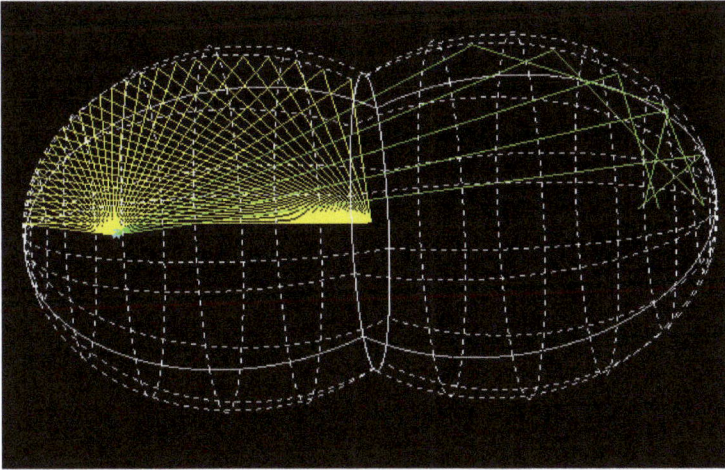

*Fig. 2.    Rays in a compound mirror that is made up of two partial ellipsoidal reflectors. There are rays that are concentrated on an intermediate focal point and there is a fan of rays that impinges on the second partial ellipsoidal reflector without passing through the intermediate focal point.*

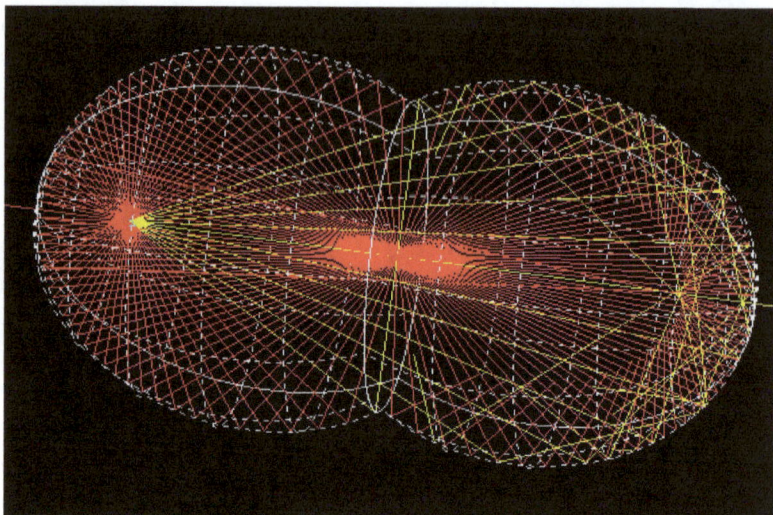

*Fig. 3.     Compound mirror made up of two ellipsoidal parts. The ray fan, the rays of which do not pass through the common focal point of the two ellipsoids, is depicted in yellow color.*

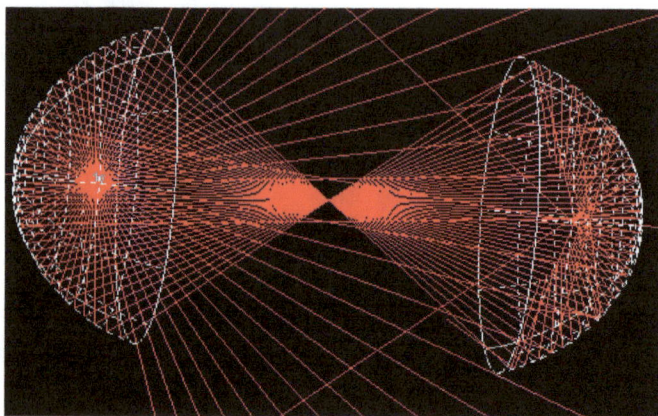

*Fig. 4.     Open reflector system consisting of two ellipsoidal surfaces. Light rays emanate from the focal point on the left-hand side.*

In Fig. 5, a symmetric mirror system is shown as illumination device, but it is also possible that both reflectors differ in size. In this way, they can be adopted to geometrical conditions of the light source and/or the region of the floating zone.

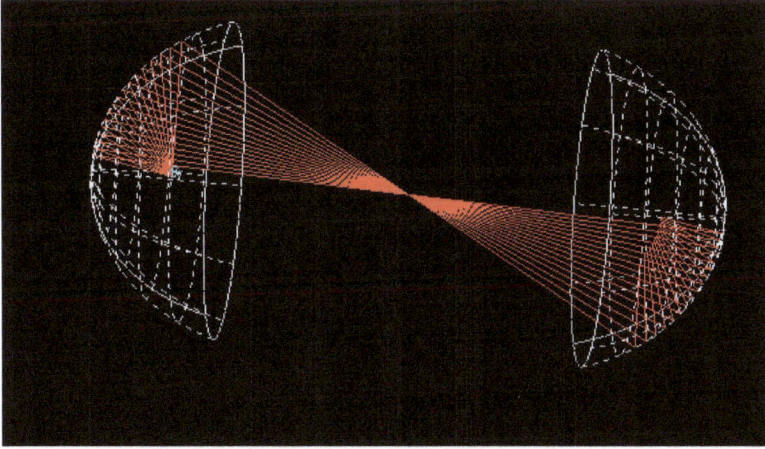

*Fig. 5.    Open mirror arrangement with intermediate focus. Both partial ellipsoid reflectors have a focal point in common.*

## 2.2    Dedicated set-up for crystal growth

Let us now have a look at an experimental arrangement that was used in laboratories for crystal growth. The optical system of this experimental arrangement [1] was developed starting from a commercial unit for cinematographic projection.

To reach the high temperatures necessary for the crystal-growth process, a carbon arc was used as intense light source. In today's experimental configurations based on the same working principle, a halogen lamp is used as source of light.

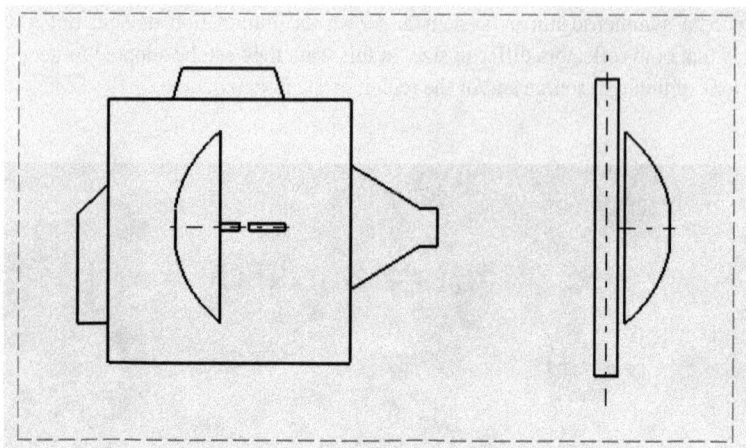

*Fig. 6.    Modification of a cinema-projection unit for the purpose of crystal growth.*

The first ellipsoid mirror collects the light form the source and re-directs it. The light rays then pass through the plane of the intermediate focal point, which might be considered as a virtual light source for the second mirror. The second reflector serves to illuminate the melt region with high intensity.

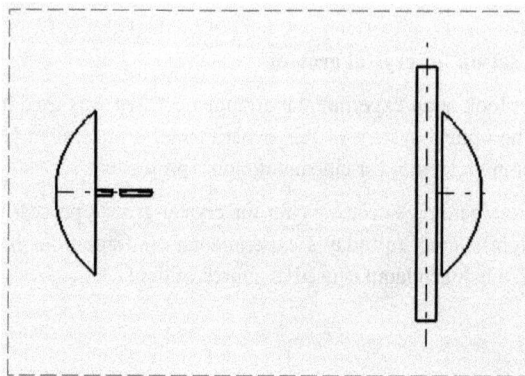

*Fig. 7.    The core part of the crystal-growth set-up that is based on a modified cinema-projection unit. A carbon arc in horizontal orientation is part of the light source. The two rods and the floating zone are in a quartz tube on the right-hand side.*

*Fig. 8.    Top view of the experimental arrangement in Fig. 7.*

This experimental set-up features a pulling mechanism to move the rods through the second focal point. In addition, the rods can be rotated in the same direction or in opposite directions, respectively. This depends on the experimental situation. Another important experimental parameter to be controlled is the speed of rotation.

The rods are situated in a tube made of quartz, which can be set under pressure.

The device was successfully used to grow single crystals of spinels and a variety of other oxides.

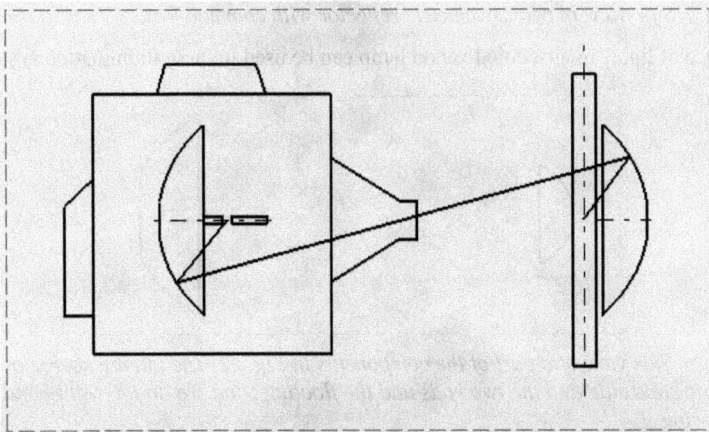

*Fig. 9.    Crystal-growth set-up based on a modified cinema-projection unit. A selected ray is shown to illustrate the working principle. The intermediate focal point is at the position where this ray crosses the axis of symmetry of the two mirrors.*

### 3. Vertical design

In other facilities for crystal research, a configuration is used that features a vertical design [2]. This creates a more symmetric illumination of the melt region, because the second mirror and the rods are oriented along a common axis. Both partial reflectors have a common focus.

In some designs, the first and the second mirror do not have the same shape and are of different size.

*Fig. 10.    Side-view of two asymmetric reflector with common focus.*

As source of light, an air-cooled xenon lamp can be used in these illumination systems.

*Fig. 11.    Side view of a part of the components in Fig. 12. The intense source of light is at the left-hand side and the two rods and the floating zone are on the right-hand side of this drawing.*

From the mechanical point of view, it seems more difficult to introduce a pulling mechanism into this experimental arrangement. To control the light flux, these optical systems are sometimes equipped with mechanical shutters in a plane between the intermediate focus and the molten zone.

Figures 12 to 14 provide three-dimensional views of key parts of the crystal-growth installation to illustrate the experimental configuration.

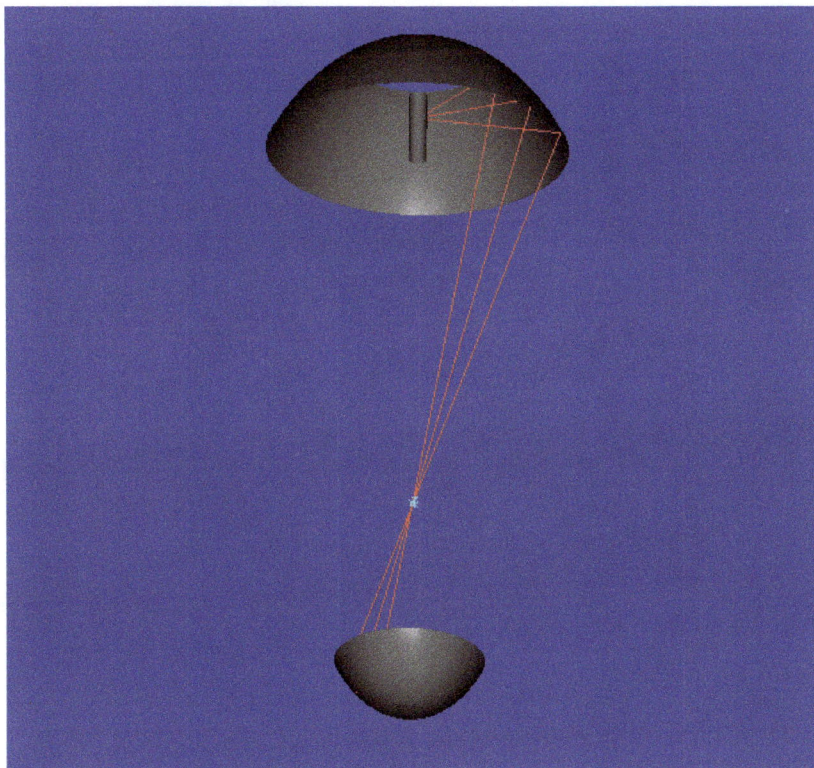

*Fig. 12.    Vertical design of an illumination system based on two asymmetric reflectors. View 1: The light rays impinge on the floating zone that is formed between the two rods.*

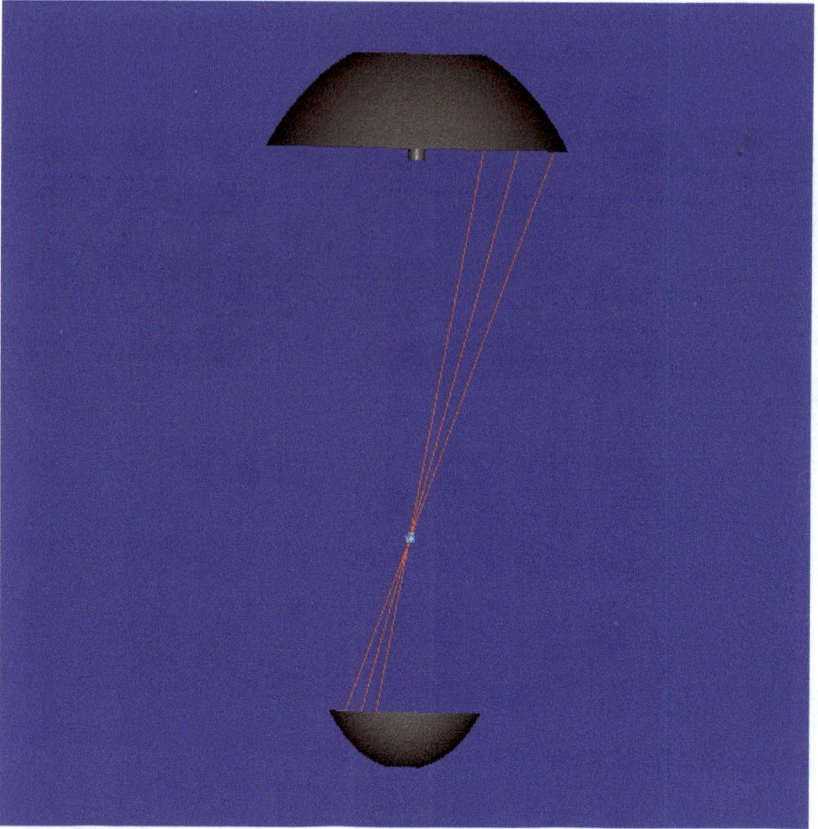

*Fig. 13.   Vertical design of an illumination system based on two asymmetric reflectors. View 2: A side-view of the experimental arrangement that brings out the position of the intermediate focal point.*

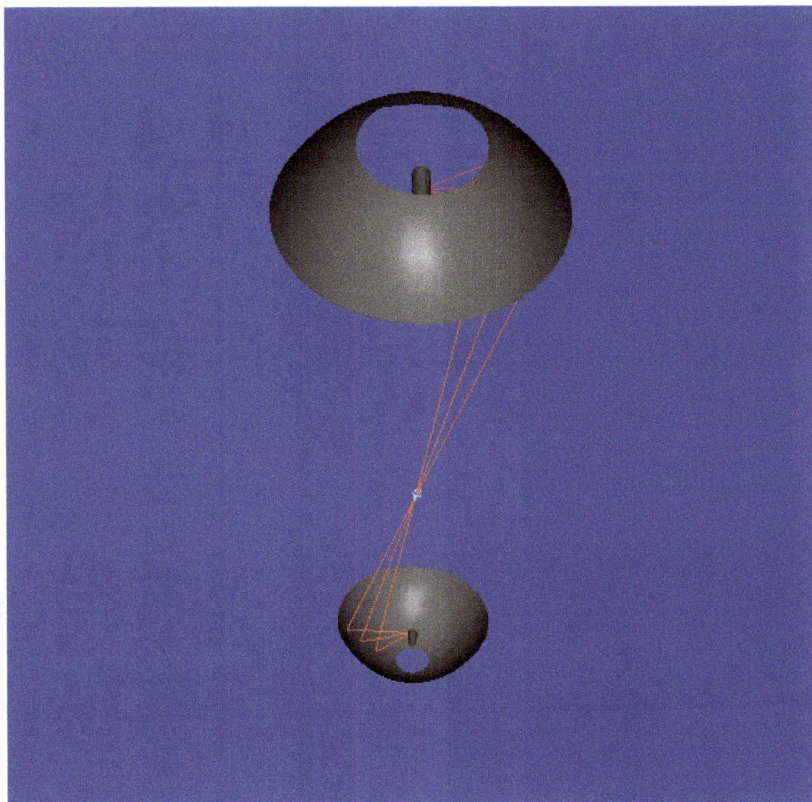

*Fig. 14.    Vertical design of an illumination system based on two asymmetric reflectors. View 3: View inside the smaller partial ellipsoidal mirror. The source of light is visible.*

**References**

[1]    C. Kooy, H.J.M. Couwenberg, Zonesmelten van oxydes in een koolboogbeeldoven, Philips Technisch Tijdschrift 5 (1961) 143-148.

[2]    H.A. Dabkowska, A.B. Dabkowski, Crystal Growth of Oxides by Optical Floating Zone Technique, in: G. Dhanaraj, K. Byrappa, V. Prasad, M. Dudley (Eds.), Springer Handbook of Crystal Growth, Springer-Verlag, Berlin, Heidelberg, 2010, pp. 367-391. https://doi.org/10.1007/978-3-540-74761-1_12

# CHAPTER 5

# Laser heating

## Abstract

A typical experimental set-up used for laser heating for crystal growth is presented. It is an example of the application of coherent light sources in the field of crystal growth. The method of laser-heated pedestal growth makes use of a $CO_2$ laser, for example, to illuminate region of the floating zone with intense radiation. Some of the optical components used in this method are shown in detail. The so-called reflaxicon is a dedicated optical device that is used to increase the diameter of the laser beam before directing it to the floating zone by a pair of annular mirrors.

## Keywords

Laser-heated pedestal growth method, annular mirrors, coherent light source, reflaxicon, $CO_2$ laser

## Contents

1.    Introduction..........................................................................................89

2.    Laser-heated pedestal growth method.................................................89

References ...............................................................................................98

# 1.    Introduction

In chapters 3 and 4, experimental configurations were discussed that made use of incoherent light sources, but coherent sources of light can also be applied to illuminate the specimen. The infrared radiation emitted by some lasers is used to melt the polycrystalline material and to reach high temperatures in the floating zone.

We will have a look at the laser-heated pedestal growth method [1,2], an established method to grow crystal by laser heating.

# 2.    Laser-heated pedestal growth method

A typical experimental configuration of the laser-heated pedestal growth method consists of the coherent light source and two annular mirrors. One of these ring-like reflectors is flat while the other one has a paraboloidal shape (Fig. 1).

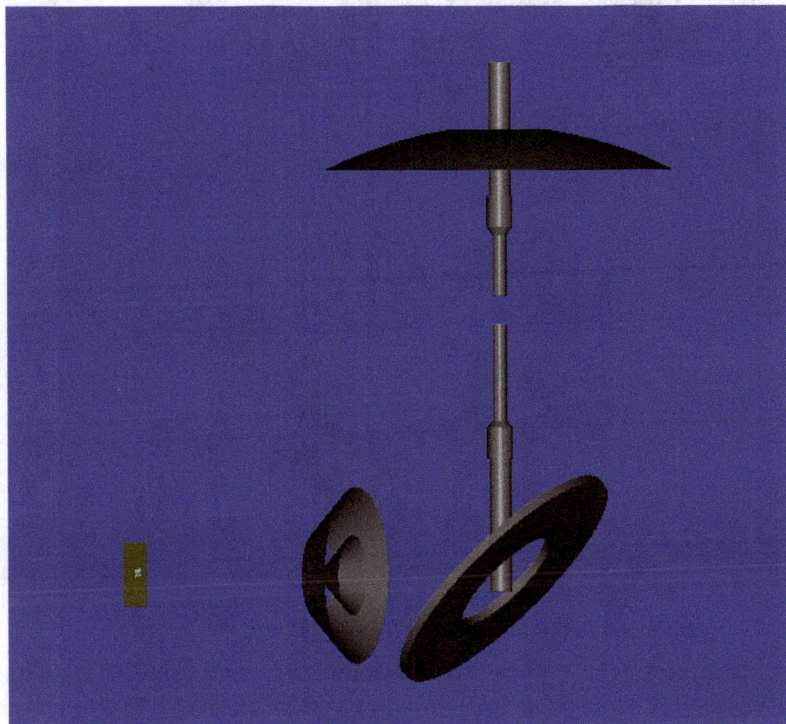

*Figure 1.    A reflector system used for laser heating.*

The symmetry axes of the paraboloidal mirror and the two rods coincide. The melting zone, which is created by the intense radiation of the laser beam, is situated between the two rods (Fig. 2).

A $CO_2$ laser can be used to provide the intense infrared radiation. Nd:YAG lasers represent another source of coherent radiation that is applied to laser heating. The crystal-growth set-ups are usually also equipped with a helium-neon laser for alignment purposes.

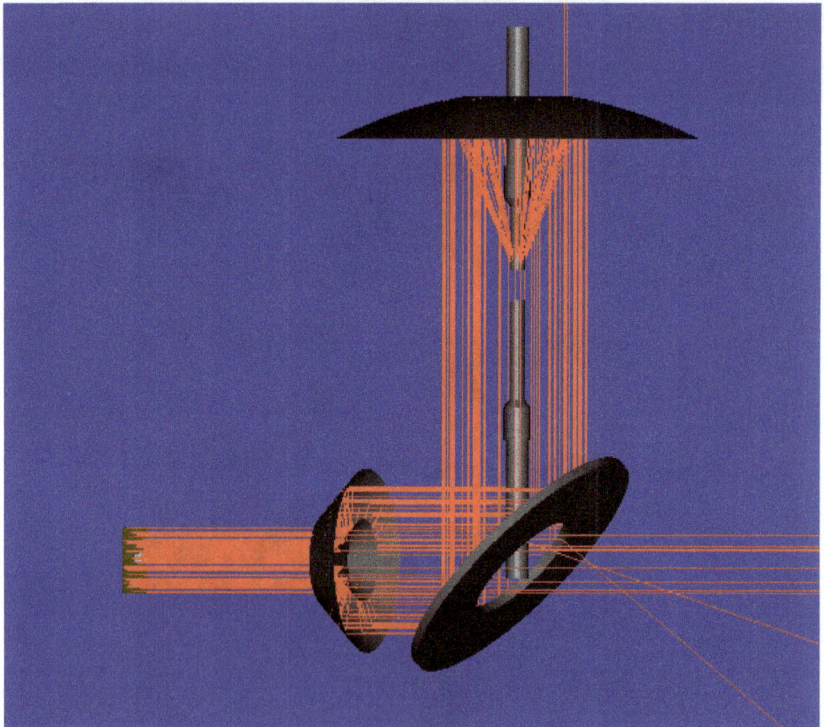

*Figure 2.     Light rays from a laser passing through the reflector system.*

An important component of the laser crystal-growth set-up is the reflaxicon [3] (Figs. 3-5). I suppose that this word is built after the model of the word "axicon". This is a refractive component with a similar function [4]. The first part of the word "reflaxicon" indicates that this optical component is based on reflection.

The purpose of the reflaxicon is to expand the collimated laser beam coming from the laser. To this end, two mirrors with circular symmetry are used, which are mounted on the same optical axis.

The first mirror serves to reflect the incoming rays by 90°, i.e. in a direction perpendicular to the optical axis (Fig. 2 and Fig. 5).

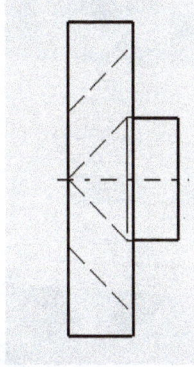

*Figure 3.    Side-view of a reflaxicon.*

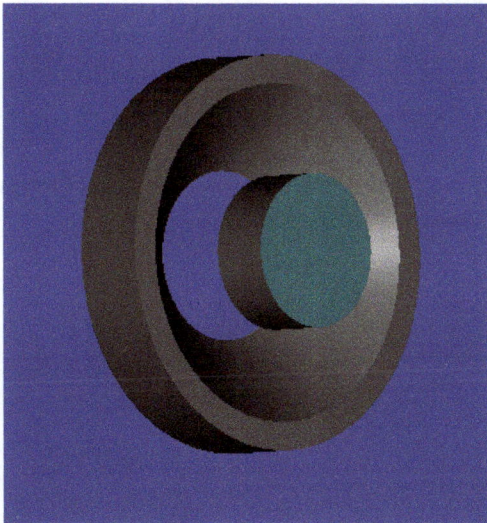

*Figure 4.    Perspective view of a reflaxicon.*

The second mirror re-directs the rays in such a way that the rays that leave the reflaxicon are again parallel to the optical axis and are also parallel with respect to each other. Fig. 5 shows the collimated beam coming out of the reflaxicon.

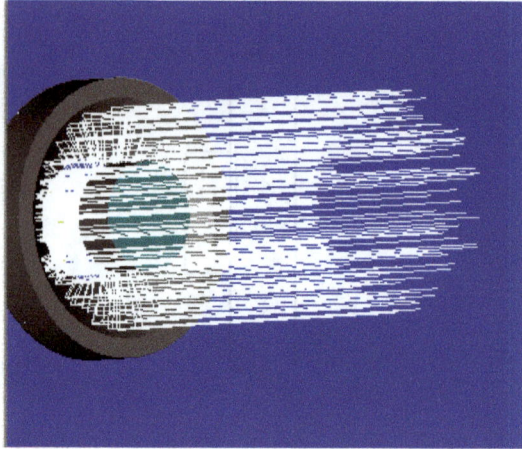

*Figure 5.     Expanded, collimated beam coming out of a reflaxicon.*

The crystal-growth set-up is equipped with a translation system that allows to move the two rods.

In Figures 6 to 10, a sequence of perspective views of the experimental configuration is provided to give a three-dimensional impression of the geometry.

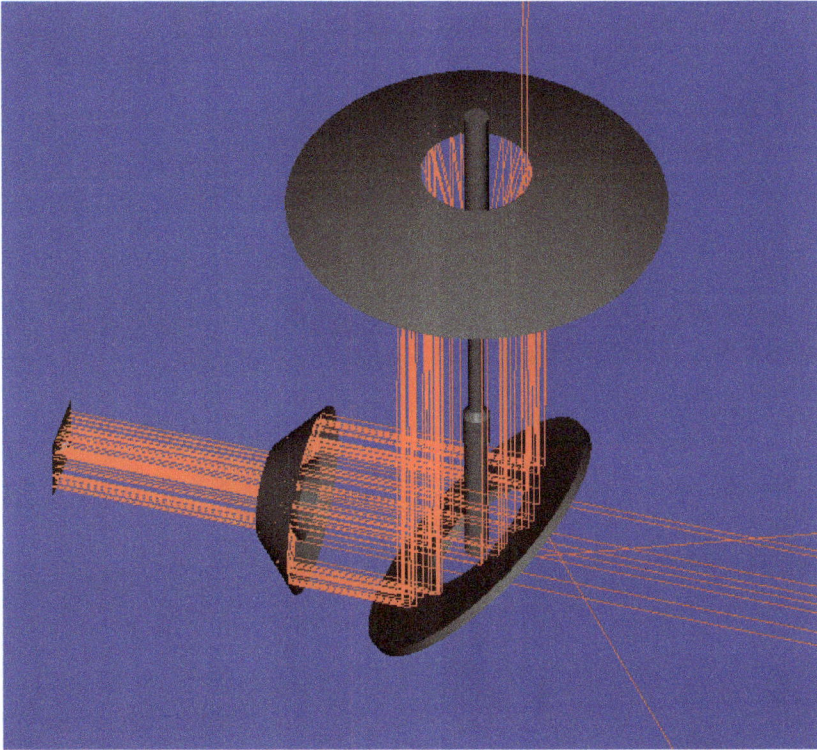

*Figure 6.*     *Laser-heated pedestal growth method. View on the first annular mirror, which is a plane reflector.*

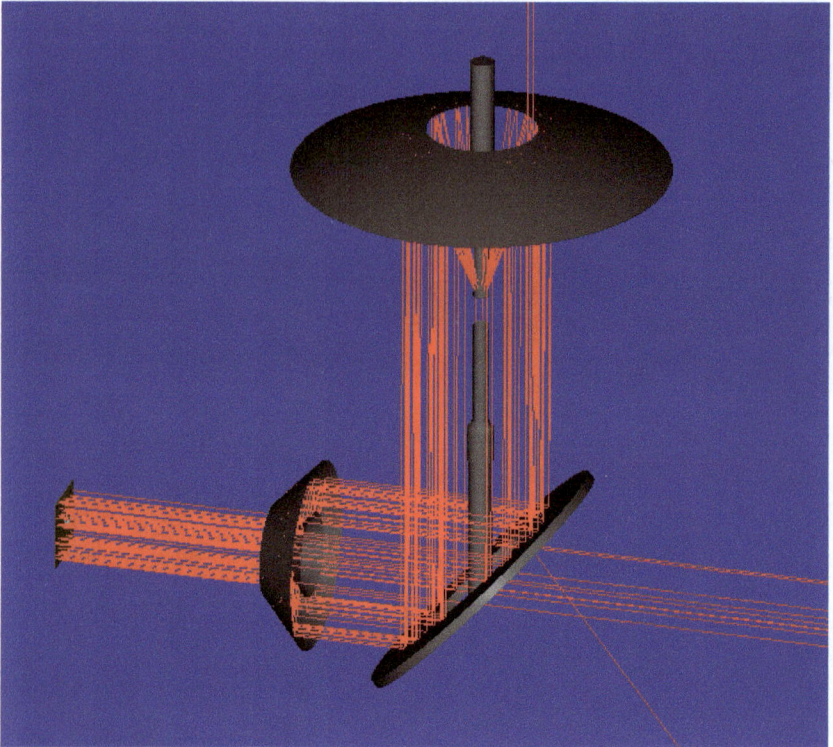

*Figure 7.     Laser-heated pedestal growth method.*

The laser is not depicted in the figures. $CO_2$ lasers and Nd:YAG lasers are in use as light sources in this method.

The development of the carbon dioxide laser dates back to 1964 [5]. Its principle wavelength bands are centered on 940 nm and 1060 nm, i.e. in the infrared region of the spectrum. It has found many applications in industry as well as in medicine.

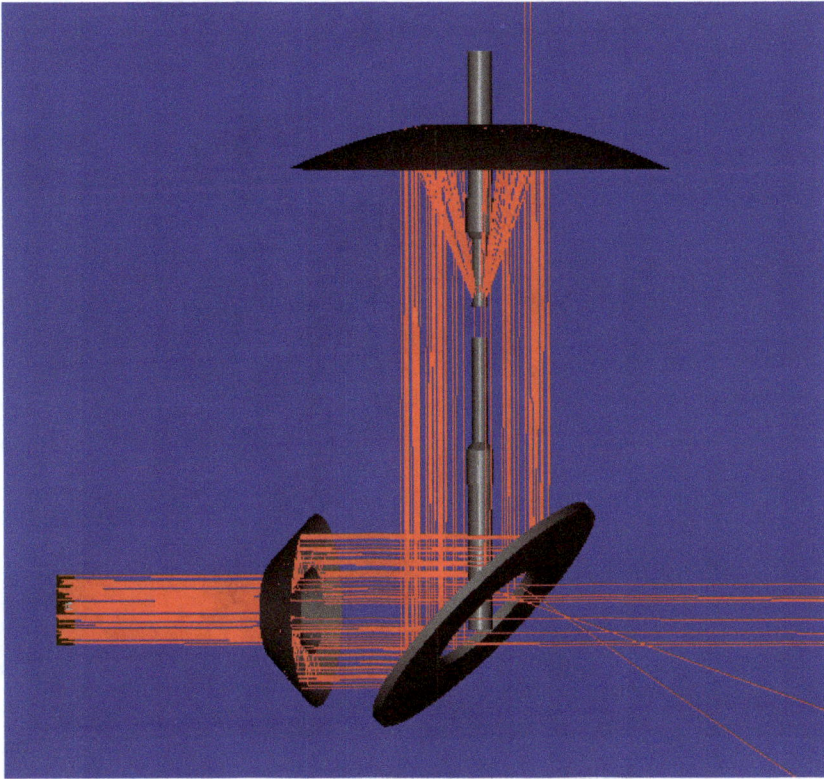

*Figure 8.    Laser-heated pedestal growth method.*

The operation of a Nd:YAG laser was demonstrated in 1964 [6]. A typical emission wavelength is at 1064 nm while other transitions are at 946 nm, 1320 nm, and 1444 nm. Descriptions of its principle of operation are given in many textbooks [7,8].

The lasing material is the crystal neodymium-doped yttrium aluminium garnet $Nd:Y_3Al_5O_{12}$. Nd(III) is used as a dopant, which replaces a small percentage of the yttrium ions in the host crystal. There is an analogy between the laser activity of neodymium ions in Nd:YAG lasers and the laser activity of chromium ions in ruby lasers.

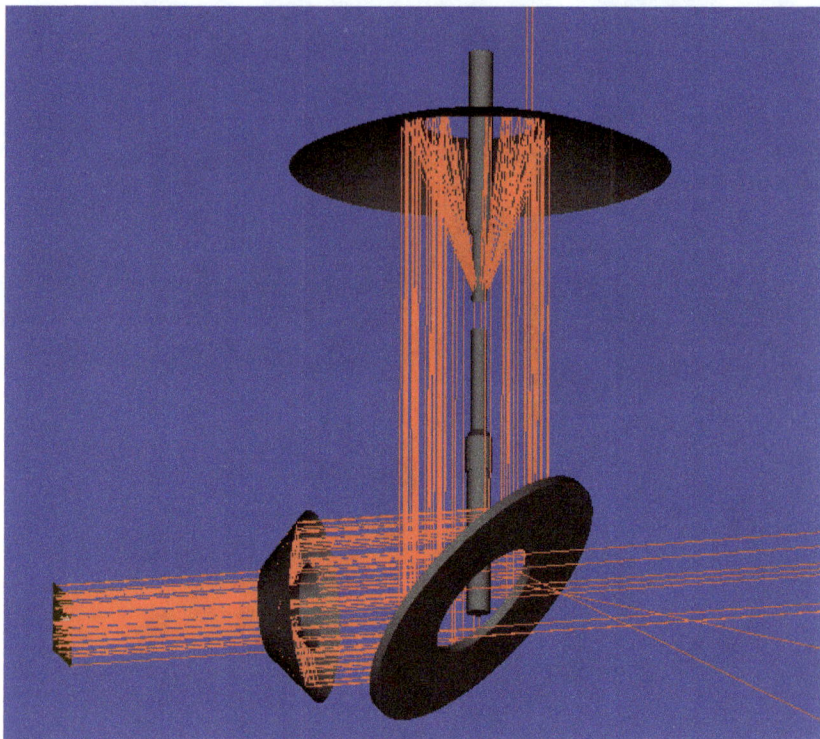

*Figure 9.    Laser-heated pedestal growth method.*

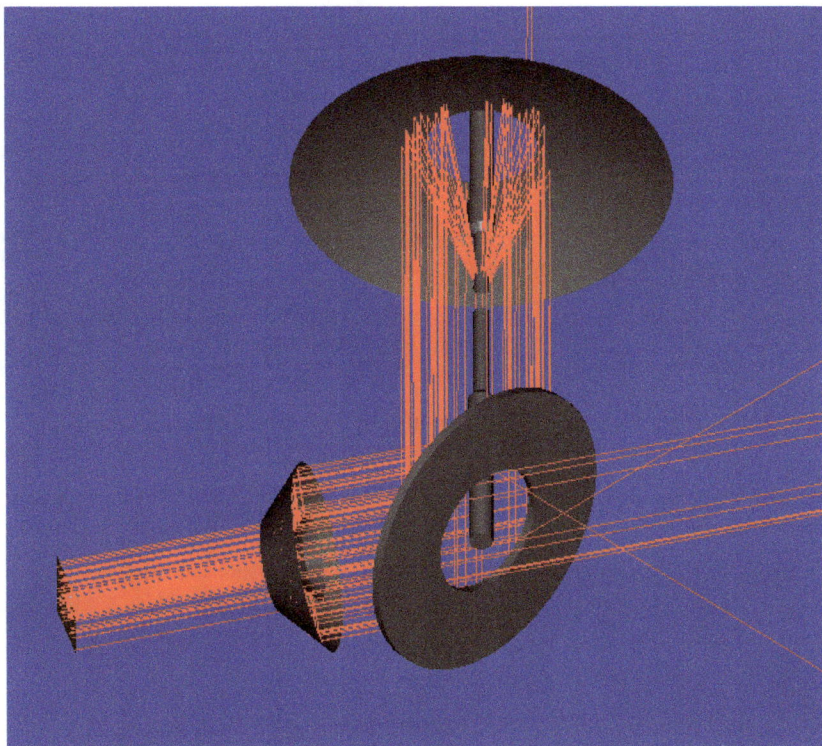

*Figure 10.     Laser-heated pedestal growth method. View inside of the second annular mirror, which is a paraboloidal reflector.*

This technique of crystal growth is used to produce single crystals of garnets and lutetia sesquioxides, among others.

**References**

[1]    M.M. Fejer, J.L. Nightingale, G.A. Magel, R.L. Byer, Laser-heated miniature pedestal growth apparatus for single-crystal optical fibers, Rev. Sci. Instrum. 55 (1984) 1791-1796. https://doi.org/10.1063/1.1137661

[2]    M.R.B. Andreeta, A.C. Hernandes, Laser-Heated Pedestal Growth of Oxide Fibers, in: G. Dhanaraj, K. Byrappa, V. Prasad, M. Dudley (Eds.), Springer Handbook of Crystal Growth, Springer-Verlag, Berlin, Heidelberg, 2010, pp. 367-391. https://doi.org/10.1007/978-3-540-74761-1_13

[3]    W.R. Edmonds, The reflaxicon, a new reflective optical element, and some applications, Appl. Opt. 12(8) (1973) 1940-1945. https://doi.org/10.1364/AO.12.001940

[4]    J.H. McLeod, The axicon, a new type of optical element, J. Opt. Soc. Am. 44(8) (1954) 592. https://doi.org/10.1364/JOSA.44.000592

[5]    C.K.N. Patel, Continuous-wave laser action on vibrational-rotational transitions of $CO_2$, Phys. Rev. 136(5A) (1964) A 1187-A 1194.

[6]    J.E. Geusic, H.M. Marcos, L.G. Van Uitert, Laser oscillations in Nd-doped yttrium aluminium, yttrium gallium and gadolinium garnets, Appl. Phys. Lett. 4(10) (1964) 182-184.

[7]    A. Yariv, Quantum Electronics, third ed., Wiley (1989).

[8]    W. Koechner, Solid-State Laser Engineering, Springer-Verlag (1992).

# Keywords

| | |
|---|---|
| Annular Mirrors | 88 |
| Aspherical Surfaces | 14 |
| Caustics | 30 |
| Closed Reflector | 30 |
| $CO_2$ Laser | 88 |
| Coherent Light Source | 88 |
| Compound Mirror | 1 |
| Compound Mirror | 30 |
| Compound Mirror | 77 |
| Conics | 14 |
| Crystal-Growth | 1 |
| Crystal-Growth Apparatus | 30 |
| Crystal-Growth Apparatus | 77 |
| Double Ellipsoid | 77 |
| Ellipses | 14 |
| Ellipsoids | 14 |
| Floating-Zone Technique | 1 |
| Floating-Zone Technique | 77 |
| Growing Techniques | 1 |
| Intermediate Focal Point | 77 |
| Laser Heating | 1 |
| Laser-Heated Pedestal Growth Method | 88 |
| Melting Points | 1 |
| Mono-Ellipsoid | 30 |
| Multiple Ellipsoids | 30 |
| Open Reflector | 30 |
| Optical Heating | 1 |
| Ray Tracing | 30 |
| Reflaxicon | 88 |
| Reflectors | 14 |

## About the Author

### Dr. Gerhard Kloos

Dr. Gerhard Kloos is a physicist working in optical engineering. He received his Ph.D. from Eindhoven University of Technology in the Netherlands. He published more than 30 papers on materials science and on optical technology. The publications on materials science cover electromechanical effects, magnetic properties, polymer physics, and photoelasticity. His interest in optical design is reflected by the books *Matrix Methods for Optical Layout* (SPIE Tutorial Text in Optical Engineering) and *Entwurf und Auslegung optischer Reflektoren* (Layout and Design of Optical Reflectors) published in German language and by patents in the field of optical technology.

He enjoys teaching optical concepts and design methods as well as trying to understand physical effects in crystals.